THE FRONTIERS COLLECTION

Springer-Verlag Berlin Heidelberg GmbH

THE FRONTIERS COLLECTION

The books in this collection are devoted to challenging and open problems at the forefront of modern physics and related disciplines, including philosophical debates. In contrast to typical research monographs, however, they strive to present their topics in a manner accessible also to scientifically literate non-specialists wishing to gain insight into the deeper implications and fascinating questions involved. Taken as a whole, the series reflects the need for a fundamental and interdisciplinary approach to modern science. It is intended to encourage scientists in all areas to ponder over important and perhaps controversial issues beyond their own speciality. Extending from quantum physics and relativity to entropy, time and consciousness – the Frontiers Collection will inspire readers to push back the frontiers of their own knowledge.

Quantum Mechanics and Gravity
By M. Sachs

Mind, Matter and Quantum Mechanics
By H. Stapp

Quantum–Classical Correspondence
By A.O. Bolivar

Quantum–Classical Analogies
By D. Dragoman and M. Dragoman

Quo Vadis Quantum Mechanics?
Edited by A. C. Elitzur, S. Dolev, N. Kolenda

Series homepage – springeronline.com

A. O. Bolivar

QUANTUM–CLASSICAL CORRESPONDENCE

Dynamical Quantization and the Classical Limit

With 14 Figures

Springer

Dr. A. O. Bolivar
Instituto Cultural Eudoro de Sousa, CEP: 72.221-970
Caixa Postal: 7316 Ceilandia, DF, Brazil email: bolivar@ime.unicamp.br

Cover figure: Image showing emerging morphological features during etching reactions, modelled by atomistic cellular automata. Courtesy of M.A. Gosalvez and R.M. Nieminen, Helsinki University of Technology.

ISSN 1612-3018
ISBN 978-3-642-05765-6

Cataloging-in-Publication Data.
Bolivar, A.O., 1972– . Quantum-classical correspondence: dynamical quantization and the classical limit/ A.O. Bolivar. p.cm. – (The frontiers collection, ISSN 1612-3018) Includes bibliographical references and index.
1. Correspondence principle (Quantum mechanics) 2. Stochastic
ISBN 978-3-642-05765-6 **ISBN 978-3-662-09649-9 (eBook)**
DOI 10.1007/978-3-662-09649-9
processes. 3. Geometric quantization. I. Title. II. Series. QC174.17.C65B65 2004 530.12–dc22 2003061038

springeronline.com

Originally published by Springer-Verlag Berlin Heidelberg New York in 2004
Softcover reprint of the hardcover 1st edition 2004

Typesetting by Stephen Lyle using a Springer TEX macro package
Final processing by Frank Herweg, Leutershausen
Cover design by KünkelLopka, Werbeagentur GmbH, Heidelberg

Printed on acid-free paper 57/3141/tr - 5 4 3 2 1 0

This book is dedicated to the memory of
Eudoro de Sousa and Mário Schönberg,
as well as our *graciosa* and *formosa* muse: Physics

Preface

In this book we address the problem of the relationship between classical and quantum physics in the non-relativistic domain. This issue involves two complementary processes:

- Given a physical system initially described by classical physics, how do we find a quantum description (quantization process)?
- Once the quantum dynamics has been obtained, how do we recover the classical physics (dequantization process)?

The concept of open system, with the isolated system as a special case, is key to our work. An open system is a physical system always interacting with its environment. The wealth of this interaction can only be apprehended through probabilistic concepts. As a model we consider the Brownian motion of a particle immersed in a fluid that has thermodynamic and hydrodynamic properties (e.g., temperature, Boltzmann's constant, coefficient of viscosity), in terms of which both the friction and diffusion coefficients occurring in the Langevin equations and the Fokker–Planck equation can be expressed.

Our quantization method (dynamical quantization) starts directly from the Fokker–Planck equation by means of a Fourier transform carrying an arbitrary parameter ℓ with the dimension of action. Quantization happens simply when we take the quantum limit $\ell \to \hbar$. We are thus able to obtain in general non-Markovian, non-Gaussian, nonlinear quantum master equations. No Hamiltonian framework is required here to quantize open systems. On the other hand the dequantization process is carried out when we take the classical limit $\hbar \to 0$ in the quantum dynamics, without making use of well-known techniques in the literature such as the WKB method, Bohm's quantum potential, decoherence mechanisms, and Ehrenfest's theorem. From our point of view, the quantum–classical correspondence stands for a quantization–dequantization process in which the Planck constant $\hbar$ plays a fundamental role during the transitions quantum domain $\to$

classical domain and classical domain → quantum domain. We imagine this physical phenomenon as being analogous to a phase transition.

At the beginning of the 20th century, the theory of Brownian movement provided an ontological status for the atomic structure of matter, thus breaking determined positivist beliefs. We hope that Brownian motion within a theory of open systems will also make it possible to regard chance as an ontological property of physics, such as space, time, and matter. Our book moves in that direction.

Acknowledgements

I thank Professor Ulrich Weiss for his hospitality at the Institut für theoretische Physik in the University of Stuttgart where local facilities such as office, computers, and libraries were very important for the writing of this book. Besides this material assistance, I wish to thank Dr. Jürgen Stockburger, Holger Baur (Dipl. Phys.), Johannes Wernz (Dipl. Phys.) for some comments and discussions about quantization of open systems without Hamiltonian and Lagrangian formalisms.

I am deeply indebted to Roman Bedau (Dipl. Phys.) for his friendship and patience in the face of my computational ignorance.

I thank Professor Daniela Dragoman (University of Bucharest, Romania) for the invitation to participate in this new collection of books from Springer-Verlag. Useful advice and assistance from Dr. Angela Lahee are also acknowledged.

In Brazil, I would like to thank Professor Sebastião Alves Dias from the Centro Brasileiro de Pesquisas Físicas (CBPF) at Rio de Janeiro, Professor Waldyr Alves Rodrigues Jr. from the Departamento de Matemática Aplicada of the Universidade Estadual de Campinas (IMECC, UNICAMP) at São Paulo, and Professor Olavo Leopoldino da Silva Filho from the Instituto de Física of the Universidade de Brasília at Brasília for reading and commenting on some parts of this book.

Financial support from the Brazilian agency Coordenação de Aperfeiçoamento de Pessoal de Nível Superior (CAPES, Contract No. 0383/02-3) is also acknowledged.

I would especially like to thank my wife Lúcia and my children Aletho and Poíesis without whom this book could not have been written. I hope their lives are also guided by the following remark from Professor Mário Schönberg

Não tenham medo, não só de levar pancada, mas também de expor suas idéias, porque se tiverem medo, nunca poderão criar algo de original. É preciso que não tenham medo de dizer alguma coisa que possa ser considerada como erro.

Do not be afraid to expose your ideas. For if you are afraid, then you will never be able to create something original. You must not be afraid of saying something that may be considered as an error.

Stuttgart
September 2003

A.O. Bolivar

Contents

1 Towards a Non-Galilean View of Physics

1.1 Isolated Systems: Galileo and Newton

Modern science was born from the notion that mathematical thought is of fundamental importance for the understanding of physical reality, in contrast with the Aristotelian world view [1]. Thus, according to Galileo (1564–1642), the true laws behind the motion of a given body can only be apprehended by neglecting or abstracting some characteristics considered as secondary, such as air resistance in experiments involving free fall of bodies, or friction for a body sliding down an inclined plane. The fact is that only when we isolate a given physical body from the world surrounding it can we give it the status of mathematical object (a material point) [2]. In this sense, a physical system whose motion we want to describe only becomes truly intelligible insofar as its essence is uniquely revealed in that ideal mathematical world. This is the ontological level of the description of physical reality.

On the basis of this Galilean mathematization, it is straightforward to confirm experimentally that the mathematical laws behind motion are never exactly satisfied, because bodies generally move in media where some kind of friction is present. Here we have the epistemological level of the description of physical reality, where the 'imperfection' of the real world becomes a mere deviation from the perfection of the ideal. In a Galilean world, a particle can be imagined as undergoing motion without any external influence, independently of the way it was started and of how it will go on. That is the content of the principle of inertia, essential for the motion of isolated systems.

Because Galilean physics is based on mathematics, it is important to bear in mind that the ontological level of the description of physical reality determines the epistemological level. For more details, the reader is referred to [1]. We represent such a relation by means of the diagram:

ONTOLOGICAL LEVEL

⇓

EPISTEMOLOGICAL LEVEL

Newton (1643–1727) in turn consolidated the Galilean program by establishing the fundamental equations of classical dynamics:

$$\frac{\mathrm{d}}{\mathrm{d}t} p = K(p, x, t) \;, \qquad \frac{\mathrm{d}}{\mathrm{d}t} x = \frac{p}{m} \;, \tag{1.1}$$

where, for the sake of notational simplicity, we shall restrict ourselves to the one-dimensional case. The generally nonlinear term $K(p, x, t)$ is a function of the position x, the momentum p, and the time t, and m is a phenomenological parameter specifying the mass of the particle. Due to the existence and uniqueness theorem for solutions to deterministic ordinary differential equations [3], the function $K(p, x, t)$ and the initial conditions $x_0 = x(t = 0)$ and $p_0 = p(t = 0)$ uniquely determine the time evolution or the trajectory of the particle.

The Function $K(p, x, t)$

The form of the function $K(p, x, t)$ has to be consistent with the concept of isolated system. To this end Newton established a kind of hierarchy for the laws of force:

1. $K = 0$ translates mathematically Galileo's principle of inertia. This absence of force determines the existence of all other forces of nature:

   Galilean absence of force $\Longrightarrow$ Newtonian forces .

2. $K = K(x)$ are forces derived from an external potential $V(x)$. Newton proposed the universality of a very important force of nature, namely, the gravitational force between any two bodies with masses m_1 and m_2, separated by a distance x:

   $$K(x) = G\frac{m_1 m_2}{x^2} \;, \tag{1.2}$$

   where G is a constant. As a consequence of restricting the forces to the particular case $K(x)$, we obtain the conservation of mechanical energy, i.e.,

   $$E = \frac{p^2}{2m} + V(x) = \text{const.} \;. \tag{1.3}$$

   Such forces are said to be conservative. Moreover, with (1.3), it turns out that Newton's equations (1.1) are endowed with symmetry under time inversion $t \to -t$. These characteristics establish (1.3) as the principle of conservation of energy.

3. $K = K(x,t)$ are forces derived from an external potential function $V(x,t)$ depending explicitly on time t. They violate the principle of conservation of energy and in general break the time-inversion symmetry. However, by adding a counter-term depending on t we can restore the required properties.
4. $K = K(x,p,t)$ are nonconservative forces. Mathematically, dissipative systems may be described by the deterministic Newton equations. For example, let us consider the case of a damped particle under the action of an external potential $V(x,t)$ and a frictional force $-\gamma p$
$$\frac{\mathrm{d}}{\mathrm{d}t}p = -\frac{\partial}{\partial x}V - \gamma p \, , \qquad \frac{\mathrm{d}}{\mathrm{d}t}x = \frac{p}{m} \, . \tag{1.4}$$
The solution with initial conditions $x(t=0) = x_0$ and $p(t=0) = p_0$ is given by
$$p(t) = p_0 \mathrm{e}^{-\gamma t} + \mathrm{e}^{-\gamma t} \int_0^t \mathrm{e}^{\gamma t'} \left[-\frac{\partial}{\partial x} V(x,t') \right] \mathrm{d}t' \, . \tag{1.5}$$
Although the friction constant γ has the same phenomenological status as the mass m, dissipative forces are not considered as legitimate physical forces, since they break the symmetry of the motion with respect to time inversion as well as violating the principle of conservation of energy. Frictional forces appear as a consequence of our inability to specify all the conservative forces generating the friction forces themselves. Here we find a situation where dynamical reversibility is a precondition for the experimentally observed irreversibility, i.e.,
$$\text{reversibility} \quad \Longrightarrow \quad \text{irreversibility} \, .$$

We wish to emphasize that, even though dissipative forces are permissible within the mathematical structure of classical dynamics, they do not fit in with the Galilean metaphysical framework based on the concept of isolated systems. In addition, it should be recognized that the universality of the gravitational force and its empirical success in describing a large class of phenomena from the motion of planets to falling apples long obscured the possible relevance of non-conservative forces within the Newtonian theoretical scheme.

Initial and Boundary Conditions

In principle, the initial position and momentum as well as the boundary conditions are mathematically determined with absolute certainty.

This is consistent with the concept of isolated system. In practice, however, it has been well-known since Galileo and Newton that there always exist some physical imperfections preventing such ideal specification. It is important to notice that Newtonian dynamics is valid for all initial conditions. Thus, recent studies of certain physical systems (e.g., the double pendulum) that are sensitive to initial conditions do not disturb the Galilean conceptual scheme. On the contrary, they presuppose its validity. These new phenomena, said to manifest conservative chaos, remain in the realm of predictability or at the epistemological level of investigation. The element of unpredictability introduces errors into the equations of motion. Hence, from a small error, the Newton equations can lead to significant effects at future times. The appearance of randomness is then merely due to our practical or computational inability to determine perfect initial conditions. We therefore encounter the paradoxical situation:

$$\text{no chaos} \quad \Longrightarrow \quad \text{chaos} \ ,$$

meaning that chaos is not an intrinsic or ontological property of classical dynamics. Rather it depends on the arbitrariness of the initial conditions.

Historically, the mathematics and physics of initial conditions (deterministic chaos) began with the account of the stability in the three-body problem in celestial mechanics (e.g., Hadamard, Duhem, Poincaré, Bruns, Birkhoff, Hénon and Heiles) and with the problem of dynamical stability in many-body systems (e.g., Maxwell, Krylov, Born) [4].

1.1.1 Hamiltonianization of Physics

With Euler, Lagrange, Hamilton, Jacobi, Poisson and others, the physics of conservative or isolated systems reached a high degree of abstraction and generality due to the introduction of the concept of generalized coordinates in terms of which the Lagrangian and Hamiltonian functions are expressed. In this formulation the main subject is no longer the force, but the potential energy (a concept from which the force may be derived) [5]. Here the Hamiltonian function is identified with the total energy of the system (1.3):

$$H(x,p) = \frac{p^2}{2m} + V(x) = E \ . \tag{1.6}$$

By means of a variational principle (the principle of least action), we obtain the Hamilton equations

$$\frac{\mathrm{d}}{\mathrm{d}t}p = -\frac{\partial}{\partial x}H(p,x) , \qquad \frac{\mathrm{d}}{\mathrm{d}t}x = \frac{\partial}{\partial p}H(p,x) . \tag{1.7}$$

These replace Newton's equations in the new formalism of classical dynamics. In the Lagrange formalism, we find the equation

$$\frac{\mathrm{d}}{\mathrm{d}t}\left(\frac{\partial L}{\partial \dot{x}}\right) = \frac{\partial L}{\partial x} . \tag{1.8}$$

This can also be generated from a variational principle in terms of the Lagrangian function L:

$$L(x,\dot{x}) = \frac{m\dot{x}^2}{2} - V(x) . \tag{1.9}$$

Once the Hamiltonian and Lagrangian structures of classical physics had been introduced, frictional forces disappeared from the theoretical framework of classical physics. Their very existence was put in doubt by the following argument: one can always imagine a dissipative system immersed in another one in such a way that the total system is conservative [6]. On the other hand, as noted by Lanczos [5], it seems to be an unimportant task to generalize the principle of least action to dissipative systems. As a consequence of the universality of the concept of isolated systems, the variational principles have played an almost magical role within physics:

> The variational principles of mechanics are firmly rooted in the soil of that great century of Liberalism which starts with Descartes and ends with the French Revolution and which has witnessed the lives of Leibniz, Spinoza, Goethe, and Johann Sebastian Bach. It is the only period of cosmic thinking in the entire history of Europe since the time of the Greeks.
>
> C. Lanczos, 1949 [5]

The enchantment, beauty and simplicity of the Hamilton and Lagrange formulations have dominated the whole of 20th century physics. We shall call this state of enchantment the principle of Hamiltonianization of physics:

> The theoretical development of the laws of motion of bodies is of such interest and importance, that it has engaged the attention of all the most eminent mathematicians, since the invention of dynamics as a mathematical science by Galileo [...]. Among the sucessors of those illustrious men, Lagrange has perhaps done more than any other analyst, to give extent and harmony

> to such deductive researches, by showing that the most varied consequences respecting the motions of systems of bodies may be derived from one radical formula [Lagrange's equations]; the beauty of the method so suiting the dignity of the results, as to make his great work a kind of scientific poem.
>
> W.R. Hamilton, 1834. Quoted in [7]

1.1.2 Quantum Physics

From the historical point of view, quantum theory appeared under the influence of ideas that arose from the Hamiltonian and Lagrangian frameworks for dealing with conservative systems. During the period of the old quantum theory (1900–1925), Planck, Bohr, Sommerfeld, Einstein, Epstein, Wilson, Schwarzschild, and others [7] introduced ad hoc rules for quantizing physical quantities such as the energy and angular momentum of the electron in a hydrogen atom. The basic tool was the action integral.

In 1925, starting from Newton's equations and replacing the physical quantities x and p with operators $\hat{x}$ and $\hat{p}$ satisfying a noncommutative algebra, Heisenberg [8] obtained equations of motion describing isolated quantum systems. One year later, from the Hamilton–Jacobi formulation of classical dynamics, Schrödinger [9] arrived at the so-called Schrödinger equation. In the same year of 1926, Dirac [10] unified the Heisenberg and Schrödinger approaches by quantizing via the Hamiltonian formalism. Finally, in 1948, Feynman [11] completed the Hamiltonianization of quantum physics by deriving the Schrödinger equation from the Lagrangian formalism of classical mechanics. As a consequence of these great discoveries, we encounter the deeply rooted belief:

$$\text{no Lagrangian and Hamiltonian} \quad \Longrightarrow \quad \text{no quantization} .$$

The Lagrangian and Hamiltonian functions are understood as preconditions for any quantization process. It is widely believed that, without them, it is impossible or unsuitable to quantize a given physical system.

The Schrödinger equation describing an isolated quantum system is given by the expression

$$-\frac{\hbar^2}{2m}\frac{\partial^2}{\partial x^2}\psi(x,t) + V(x)\psi(x,t) = \mathrm{i}\hbar\frac{\partial}{\partial t}\psi(x,t) . \tag{1.10}$$

This is a deterministic partial differential equation for the complex-valued function $\psi(x,t)$, in the sense that, given the initial condition $\psi_0 = \psi(x,t=0)$, the equation of motion (1.10) is deterministically solved for any $\psi(x,t)$. The constant $\hbar$ (Planck's constant divided by 2π) is the signature of the quantum nature of matter. The connection with experiment is made by interpreting $|\psi|^2$ as a probability density function from which average values of physical quantities can be evaluated. This is Born's statistical interpretation, responsible for the enormous empirical success of quantum physics. However, the conceptual problems related to the Schrödinger function have been the subject of unending controversy, raising questions such as:

- What is the true physical significance of the Schrödinger function ψ?
- Does the transition $\psi \Rightarrow |\psi|^2$ have any physical meaning?

The first attempts at answering those questions aimed at eliminating the ontological level of physical reality. According to Heisenberg, Bohr, Born and Pauli there is no concept of trajectory in the quantum domain. The function ψ does not describe ontological atributes such as the trajectory of an electron. Only after observing the physical property 'position in time' through a measuring device constructed for this purpose can we speak of the trajectory concept [12]: "Die Bahn entsteht erst dadurch, dass wir sie beobachten." ("The trajectory concept arises provided that we observe it.") The impossibility of any ontology in the quantum world is reflected mathematically by the Heisenberg uncertainty relations:

$$\sigma_x \sigma_p \geq \frac{\hbar}{2} , \tag{1.11}$$

where $\sigma_u = \sqrt{\langle u^2 \rangle - \langle u \rangle^2}$ for $u = x, p$.

As a consequence of establishing a physical theory without an ontological level, various epistemological interpretations of quantum mechanics have been developed, such as the Copenhagen interpretation, the many-worlds interpretation, the histories approach, and others. An unavoidable feature of such interpretations is their subjectivist character [13]:

- The esoteric ego of von Neumann, whereby the transition from ψ to $|\psi|^2$, the so-called 'collapse of the wavefunction', occurs through the special role of the consciousness of the observer carrying out a measurement.

- Human consciousness and the world possess different logical frameworks.
- The spirit of the observer or its subjectivity could be essential when describing the physical state of quantum systems. Hence physics turns out to be closely linked to psychology, neurophysiology and psychiatry.

Other eminent physicists such as Einstein, de Broglie and Schrödinger never accepted the elimination of ontology from the conceptual structure of physics. Indeed, such a claim is not even possible. According to Einstein, it would be like trying to breathe in empty space [2]. A true physical theory must be built upon an ontological basis:

> Aber vom prinzipiellen Standpunkt aus ist es ganz falsch, eine Theorie nur auf beobachtbare Grössen gründen zu wollen. Denn es ist ja in Wirklichkeit genau umgekehrt. Erst die Theorie entscheidet darüber, was man beobachten kann.
>
> From the standpoint of basic principles, it is entirely wrong to establish a theory on the basis of observable quantities alone. Indeed, the opposite actually occurs: at the outset, theory decides about what can be observed.
>
> Einstein, quoted by Heisenberg in [14], p.92

In 1952, Bohm [15] began to develop an ontological interpretation of the Schrödinger equation. Here a quantum particle actually follows a given trajectory guided by the wave ψ, independently of any measurement designed to observe it. However, just as happens in the epistemological interpretations, we should point out that there are great difficulties in establishing quantum physics on Bohm's ontological basis (for more details, see [13]). As an example of these shortcomings, we wish to mention the problem of observing chaos in the quantum domain. According to the strictly epistemological point of view of quantum mechanics, chaos should appear as a property of the measurement apparatus, provided we remain within an objective perspective. On the other hand, we can also imagine chaos as being a state of consciousness inherent in the mind of an observer performing a given measurement (subjective perspective). However, in the ontological interpretation, by definition, Bohm's quantum trajectories are not chaotic. This property is an atribute of the classical trajectories defined on the epistemological level. The Schrödinger function has no chaoticity property per se.

It is important to notice that in both the epistemological (Heisenberg, Bohr) and ontological (Bohm, de Broglie) versions of quantum mechanics, the Schrödinger equation appears as the highest achievement of the Galilean program based on the concept of isolated system. Just as the principle of inertia is the precondition for the Newtonian concept of force, the Schrödinger function ψ is a precondition for the concept of probability:

$$\text{no probability} \quad \Longrightarrow \quad \text{probability} \, .$$

Probability belongs to the epistemological level of physics. It is a concept which is therefore associated with acts of observation performed in the classical domain.

As a consequence of the Galilean view encompassing both quantum and classical physics, physical properties defined on the epistemological level do possess a minor status within the theory. Thus, chaos, irreversibility, and probability are concepts of some practical relevance. They are nevertheless absent from the true theoretical foundations of physics.

We list below three arguments suggesting the need for a non-Galilean view of physics:

- As already mentioned, when the Galilean program is extended to quantum physics, insurmountable conceptual problems arise. (For a critical and deep account of the conceptual difficulties generated by the concept of isolated systems inherent in both the epistemological and ontological interpretations of quantum mechanics, we strongly recommend the book by Grib and Rodrigues [13].) These difficulties seem to indicate that the concept of isolated system, and consequently attempts to Hamiltonianize physics, must be revised in order to achieve a conceptual harmony between classical and quantum physics.
- In recent times, provided that physical systems are open or non-isolated, i.e., they are in interaction with their surroundings, new concepts such as self-organization, dissipative structures, synergetics, fractals, and dissipative chaos have arisen showing that classical physics is not conceptually sterile. The remarkable point is that chance lies at the origin of such phenomena. This seems to indicate that the concept of isolated system, in which chance has no place, is inadequate to reveal the wealth of the physical world. Consequently, the main issue turns out to be the following: can chance be localized on the ontological level in physics? If it cannot, then what prohibits such an achievement?

- Galilean mathematization is not unique. We have to explore all possible mathematization processes, including the mathematization of chance, in order to gauge the power of this method and at the same time to reveal its limits, restrictions and weaknesses in apprehending every essence of reality. (These features are hidden by the Galilean mathematization or, in its more modern version, by the principle of Hamiltonianization of physics, whereby every source of stochasticity should be experimentally controlled on the one hand, and theoretically justified on the other.) The idea of attributing an ontological role to chance within physics seems to presuppose the existence of a non-Galilean metaphysics. (The strong influence of the Platonic philosophy upon the thinking of Galileo, the father of classical physics, and Heisenberg, the father of quantum physics [14] is remarkable.)

Considering the three arguments quoted here, we ask the specific question: is it possible to set up a non-Galilean (or non-Hamiltonian) physics based ontologically on the concept of open systems (chance), in such a way that isolated systems (without chance) are special cases? The search for an affirmative answer to this question is the main leitmotiv of the present book.

1.2 Open Systems

An open system is one which is always in interaction with its environment. When we drop the environment, we obtain an isolated system as a special case. At least two interesting questions can be given a meaning in this approach:

- Under what physical conditions can we isolate a given system?
- For what types of physical system is the isolation process possible?

As a prototype of an open or non-isolated physical system, we imagine a Brownian particle immersed in a fluid. The construction of a theory of open systems is not a trivial task. Such a theory must possess at least four fundamental properties: irreversibility, synergeticity, fractality and chaoticity.

1.2.1 Prigogine: Irreversibility

Due to the interaction process, our Brownian particle dissipates energy to the environment in an irreversible way. On the other hand, the influence of the environment is responsible for activating the motion of the Brownian particle, causing fluctuations in the physical quantities describing it (e.g., position, momentum). The interchange between dissipation and fluctuation is a characteristic feature of open systems. The problem is to describe such an interchange. More specifically, in order to establish a complete theory of open systems, we must account for the way in which the stochastic physics of the environment influences the dynamics of our Brownian particle.

The dynamics of open systems is conditioned by boundary conditions, which are in turn determined by the physics of the environment. We thus assume that this physics is a kind of thermo-hydrodynamics, so that the diffusion and friction coefficients present in the equations of motion of the Brownian particle can be expressed in terms of thermodynamic and hydrodynamic quantities (e.g., the viscosity of a fluid, temperature, Boltzmann constant) (see Fig. 1.1).

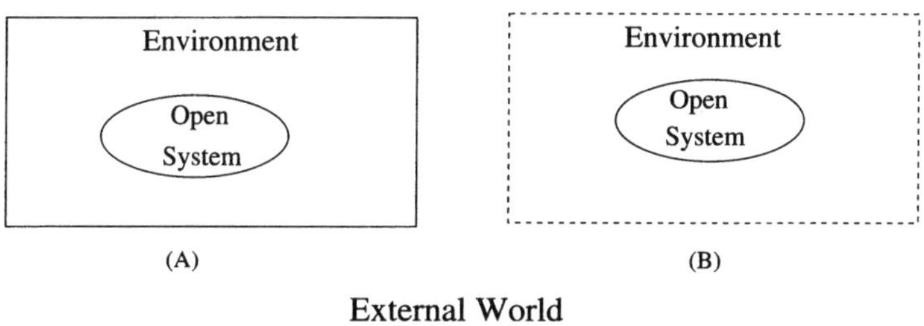

Fig. 1.1. The system is said to be dynamically open insofar as it is immersed in an environment which may in turn be described by the thermodynamics of equilibrium (**A**) or nonequilibrium (**B**)

From the viewpoint of thermodynamics, our system as a whole (Brownian particle plus environment) may be characterized by properties of thermodynamic equilibrium and in more common cases by nonequilibrium states. In both contexts the Boltzmann constant k_B plays a fundamental role. It corresponds to the more general signature of the thermodynamic nature of the environment. When the environment is a thermal reservoir or a heat bath, the temperature turns out to be an important physical quantity characterizing the equilibrium thermodynamics. Otherwise, nonequilibrium thermodynamics must be invoked.

A thermodynamic system can be classified according to its relation with the external world (see Fig. 1.2):

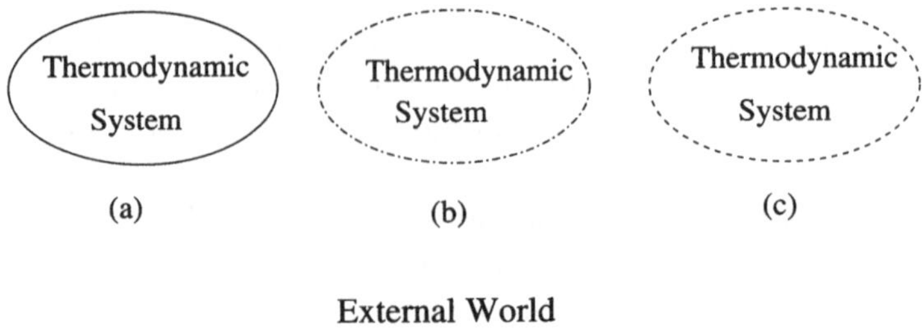

Fig. 1.2. From the thermodynamical point of view, the total system of Fig. 1.1 (open system plus environment) may be considered as isolated (**a**), closed (**b**), or open (**c**) with regard to the external world

- isolated system: exchanges neither energy nor matter with the external world,
- closed system: exchanges energy but not matter with the external world,
- open system: exchanges both energy and matter with the external world.

The change of entropy is responsible for the reversible and irreversible properties:

$$\mathrm{d}S = \mathrm{d_e}S + \mathrm{d_i}S \ ,$$

where $\mathrm{d_e}S$ is the flow of entropy due to interactions with the external world, and $\mathrm{d_i}S$ the changes of entropy within the system. When $\mathrm{d_i}S = 0$ we have a reversible thermodynamic process, whereas $\mathrm{d_i}S > 0$ indicates an irreversible process. For isolated systems we find

$$\mathrm{d}S = \mathrm{d_i}S \geq 0 \ .$$

In the domain of nonequilibrium thermodynamics, Prigogine's theory of irreversibility [16–20] has revealed novel features underlying the physics of thermodynamically open systems. This theory investigates dissipative structures, e.g., Bénard instability in hydrodynamics (Fig. 1.3), thermodiffusion (Fig. 1.4), and self-organization processes in far-from equilibrium situations caused by fluctuation phenomena. We do not intend to examine this theory. We wish only to emphasize the fundamental relevance of the concept of irreversibility for the definition of dynamically open systems, such as our Brownian particle.

1.2.2 Haken: Synergetics

Another property of open systems is based on the synergetic principle due to Haken [21–23]. This says that the dynamic behavior of each

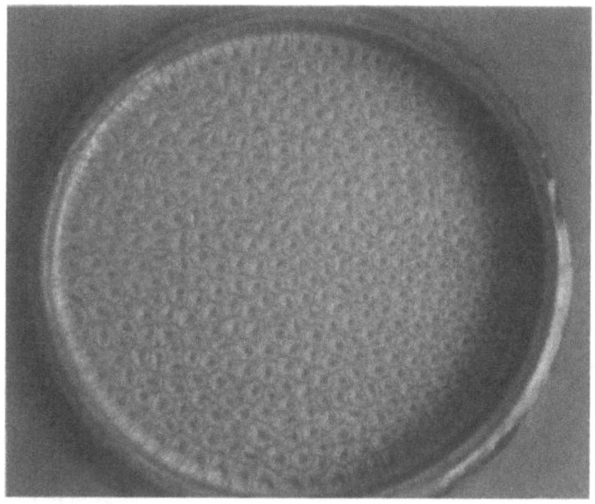

Fig. 1.3. Bénard cells are an example of dissipative structure arising in situations of thermodynamic nonequilibrium and hydrodynamic instability

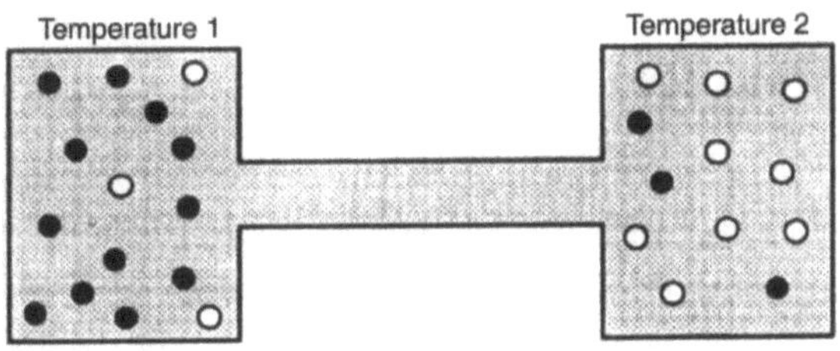

Fig. 1.4. Phenomenon of thermodiffusion. Consider a mixture of gases (e.g., hydrogen and hydrogen sulphide). Due to a temperature difference between the two vessels, a gradual separation of the gases occurs. The hotter vessel is richer in the lighter hydrogen gas, whereas the cooler vessel has a higher concentration of the more massive hydrogen sulphide

environment particle cooperates in a stochastic way, through frictional collisions for instance, bringing about the stochasticity of the Brownian motion. All these stochastic features are encapsulated in the friction and diffusion coefficients present in the equations of motion. The most characteristic example of a cooperative phenomenon is the laser (see Fig. 1.5).

This principle is essential for the definition of open systems. When the cooperation starts to weaken, the motion of each particle can be approximately considered as independent.

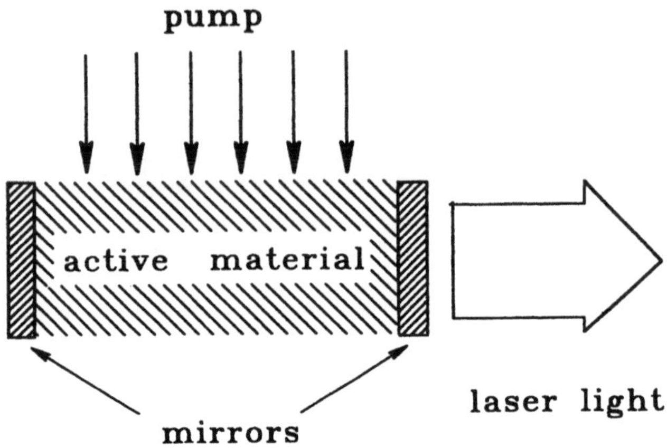

Fig. 1.5. An active material device with two mirrors at its end faces is energetically pumped from the external world to produce laser light

1.2.3 Mandelbrot: Fractals

Geometrically, as a consequence of the environmental dynamics, the Brownian trajectories present an irregular or fragmented form. These curves are characterized mathematically using the concept of fractal dimension. In more technical terms, a fractal is defined as being a set for which the Hausdorff–Besicovich dimension strictly exceeds the Euclidean or topological dimension (Fig. 1.6). For a certain kind of Brownian motion (the Wiener process), the fractal dimension is 2, while the topological dimension is 1 [24].

The geometric framework of open systems is fractal geometry. For isolated systems, we lose the fractal features and recover the Euclidean ones. Chance therefore becomes the source of the fractal geometry of nature [24]: "The study of fractals seems [...] to increase one's understanding of randomness."

1.2.4 Lorenz and Ruelle: Dissipative Chaos

We conceive of the initial conditions as stochastic variables having a given probability distribution function. Within a theory of open systems, the deterministic initial conditions appear as a special case among several other possible conditions. An intrinsic feature of the Brownian trajectories is that they are sensitive to initial conditions, a dynamical property known as chaos.

The most significant development in chaology has arisen in the context of nonconservative systems, in which friction forces play a funda-

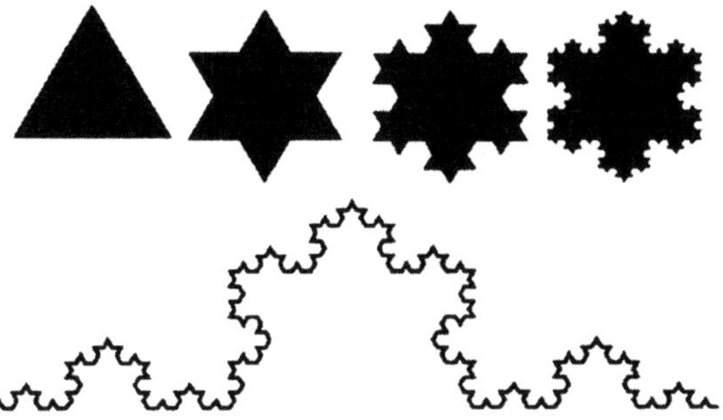

Fig. 1.6. A Koch curve constructed from successively reduced triangles. Its fractal dimension is approximately 1.28

mental role, such as in atmospheric convection [25,26], and in turbulence [27,28]. In this context of dissipative chaos, new mathematical concepts have been formulated, e.g., the strange attractor (Fig. 1.7), bifurcations, the Lyapunov exponent, and others [29]. It is important to note that only within a theory of open systems do both concepts of dissipative and conservative chaos acquire an ontological status in physics. Indeed, non-chaotic or regular initial conditions form only a small and very special class amongst them.

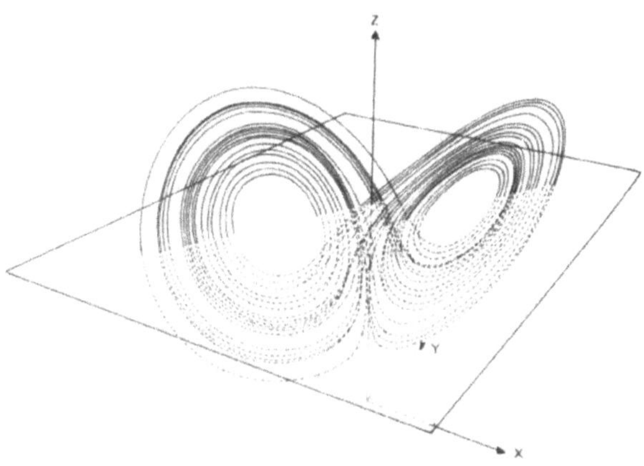

Fig. 1.7. Lorenz strange attractor generated by dissipative deterministic equations of motion

An open and stimulating issue is to establish a theory of stochastic chaos for which deterministic chaos arises as a special case. Relevant advances in this field have been achieved by Arnold [30].

On the basis of this brief sketch of the properties of open systems, we make the following remarks:

- Open systems are primary, while isolated systems are treated as special cases.
- The fundamental ingredient generating the concept of an open system is chance.
- The mathematization of chance is the key concept absent from the Galilean view of physics.
- According to a non-Galilean mathematization of physical reality, chance reveals itself as a source of new concepts in physics as well as in mathematics.

1.2.5 Our Model: Brownian Motion

The existence of the environment (a fluid) causes a given particle to undergo an extremely irregular movement called Brownian motion. We suggest placing the birth of the open system in 1784 when the Dutch physicist Jan Inghen-Housz [31] first glimpsed Brownian motion under the microscope:

> I have often troubled my head about the problem of finding a method to avoid the too rapid evaporation of a drop of water, or any other liquid, in which I wanted to observe insects [...]. Even if one wishes to observe the shape and size of some of these corpuscles for even the short time during which such a droplet lasts in the focal point of a microscope, one must agree that, as long as the droplet lasts, the entire liquid and consequently everything which is contained in it is kept in continuous motion by the evaporation, and this motion can give the impression that some of the corpuscules are living, even if they have not the slightest life in them. To see clearly how one can deceive one's mind on this point if one is not careful, one only has to place a drop of alcohol in the focal point of a microscope and introduce a little finely ground charcoal therein, and one will see these corpuscules in a confused, continuous and violent motion, as if they were animalcules which move rapidly around.
>
> J. Inghen-Housz, 1784. Translation in [32]

In 1827, a more detailed account of this phenomenon was given by the Scottish botanist Robert Brown [33]. He investigated granules of a plant called *Clarkia pulchella* under the microscope [34]:

> While examining the form of these particles immersed in water, I observed many of them very evidently in motion [...]. These motions were such as to satisfy me, after frequently repeated observtion, that they arose neither from currents in the fluid, nor from its gradual evaporation, but belonged to the particle itself.
>
> R. Brown, 1828. Quoted in [34]

A more modern description of Brownian motion is given by Jean Perrin [35]:

> In a fluid mass in equilibrium, such as water in a glass, all the parts appear completely motionless. If we put into it an object of greater density, it falls. The fall, it is true, is the slower the smaller the object; but a visible object always ends at the bottom of the vessel and does not tend again to rise. However, it would be difficult to examine for long a preparation of very fine particles in a liquid without observing a perfectly irregular motion. They go, stop, start again, mount, descend, mount again, without in the least tending toward immobility.
>
> J. Perrin, 1909. Quoted in [24]

The first mathematization of Brownian motion was elaborated by Einstein in 1905 [36,37]. He built a theoretical model in which the environment was assumed to be a liquid in thermodynamic equilibrium. The liquid particles act upon the Brownian particle in a probabilistic way. Einstein then derived an equation of motion for the probability density function $F(x,t)$ in terms of the stochastic position X alone:

$$\frac{\partial}{\partial x}F(x,t) = D\frac{\partial^2}{\partial x^2}F(x,t) \; . \tag{1.12}$$

This allowed him to predict the following result for the fluctuation in position:

$$\langle (X - \langle X \rangle)^2 \rangle = 2Dt \; , \tag{1.13}$$

where D is a diffusion constant containing some geometrical, mechanical and thermodynamic properties, such as the size of the Brownian particle, the friction constant, the Boltzmann constant and the temperature of the liquid. Some years later this result was confirmed experimentally by Perrrin [35].

In 1908, Langevin [38] proposed an extension of Newton's equations (1.1):

$$\frac{\mathrm{d}}{\mathrm{d}t}P = \gamma\frac{P}{m} + b\Phi\,, \qquad \frac{\mathrm{d}}{\mathrm{d}t}X = \frac{P}{m}\,, \tag{1.14}$$

where X and P can be interpreted as stochastic processes due to the existence of the fluctuating force Φ. In the Langevin approach, Brownian motion is seen explicitly as a dissipation and fluctuation phenomenon. [We shall see in more detail in Chap. 3 that the Langevin equations (1.14) reduce to the conservative Newtonian equations when both the friction and fluctuation constants γ and b go to zero.] Using some statistical considerations concerning certain properties of the force Φ, Langevin [38] also derived Einstein's formula (1.13).

In the physics of isolated systems, the concept of probability has only an epistemological status. Hence Brownian motion, despite its valuable practical achievements (e.g., the measurement of Boltzmann's constant), never received an ontological status in physics. On the one hand, Einstein never formulated a relativistic theory of Brownian motion. On the other hand, even though it was closely related to the atomic structure of matter, Planck, Bohr, Heisenberg, Born, Schrödinger, and others never attempted to obtain a quantum theory of Brownian motion. Moreover, Einstein imagined the task of physics as being to justify the epistemological importance of the probability concept from a theory constructed ontologically on the grounds of a theory of isolated systems. (The ergodic theory, in which statistical properties are meant to be derived from the dynamics of isolated systems, was a model for Einstein [39].) In this view, Brownian motion should always be reduced to the statistical physics of isolated systems. For example, one glorious task has been to elaborate Hamiltonian models from which one can derive the Langevin equations (1.14).

In order to contribute to a general theory of open systems and bring out the ontological status of Brownian motion in physics, the specific aim of the present book is to address the following question: how can we investigate Brownian motion in quantum physics without using Hamiltonians and Lagrangians? Recently we have worked out a method for quantizing open systems without using Hamiltonian and Lagrangian formalisms [40–44]. Furthermore, we have also implemented an alternative procedure for calculating the classical limit $\hbar \to 0$ of conservative and nonconservative quantum dynamics [41,42,44–47]. The present monograph aims to provide more detail concerning the conceptual consequences of our work with regard to the relationship between quantum and classical physics.

The book is organized as follows. Since open systems are intrinsically probabilistic phenomena, Chap. 2 presents some concepts from the theory of stochastic processes, such as the probability space and stochastic variables. Chapter 3 deals with the classical physics of dissipation and fluctuation phenomena. As model, we use Brownian motion in configuration space (Sect. 3.2) as well as in phase space (Sect. 3.3). When we isolate the system in Sect. 3.4, we obtain the deterministic equations as special cases. In Sect. 3.5, we present the program of Hamiltonianizing physics, i.e., studying an open system as an isolated system (the Caldeira–Leggett approach).

In Chap. 4, we discuss the weak points of the historically accepted quantization methods based on the existence of Lagrangian and Hamiltonian formalisms. These methods elaborated by Planck and Bohr (Sect. 4.2), Heisenberg (Sect. 4.3), Schrödinger (Sect. 4.4), Dirac (Sect. 4.5), and Feynman (Sect. 4.6) work well for a determined class of conservative systems and are in general ambiguous for nonconservative systems. The first alternative quantization method for deriving the Schrödinger equation is mentioned in Sect. 4.7 (Nelson quantization). Of special interest is the quantization scheme of Sect. 4.8 based on the concept of entropy (Olavo quantization). In Sect. 4.9, we quantize open systems without Hamiltonians and Lagrangians. Since we start directly from the equations of motion, we call this method dynamical quantization. As a specific example, we quantize a type of Brownian motion. In this way, we are able to derive the Caldeira–Leggett master equation, valid for a particular class of stochastic processes, the so-called Markovian processes. Moreover, our quantization method also leads to non-Markovian quantum master equations. When we isolate the system, an interesting consequence of this novel quantization procedure is that the Schrödinger function (or wavefunction) is derived from the von Neumann function (or density matrix). This contrasts with the procedure usually followed in the textbooks, whereby one derives the density matrix from the wavefunction.

Chapter 5 approaches the problem of the transition from quantum mechanics to classical mechanics. The Bohr correspondence principle (Sect. 5.2), the Ehrenfest theorem (Sect. 5.3), the WKB Method (Sect. 5.4), the Feynman method (Sect. 5.5), the decoherence approach (Sect. 5.6), and the Bohm quantum potential method (Sect. 5.7) are criticized. In Sect. 5.8, in order to establish the logical consistency of the dynamical quantization method, we define an alternative procedure for calculating the classical limit of the quantum dynamical equations for both conservative and nonconservative systems. In the last chapter, we sum up the main results of our work and emphasize some open questions.

There are five appendices. Appendix A deals with some notions of set theory essential for understanding the probability concept. In Appendix B, we quantize the Smoluchowski equation. Appendices C and D are devoted to a comparison between the Dirac and Feynman quantizations and dynamical quantization for isolated systems. Finally, in Appendix E, we present the Wigner representation, in which we derive an operator structure for classical mechanics. In addition, we maintain that this new representation of classical mechanics is a more suitable route for quantizing physical systems independently of the existence of Lagrangians and Hamiltonians.

2 Some Basics of Stochastic Processes

2.1 Motivation: Open Systems as Stochastic Processes

The basic characteristic of the mathematical thinking here is to reveal reality on the following two levels [1]:

ONTOLOGICAL LEVEL
$\Downarrow$
EPISTEMOLOGICAL LEVEL

The process of mathematizing chance illustrates this scheme. (More historical details about the mathematization of chance can be found in [?].) Let us start from the somewhat domineering Laplacian method for eliminating any ontological basis for the concept of probability. On the basis of the Newtonian physics of isolated systems, Laplace created a metaphysical being endowed with an intelligence such that for him all the details of nature are always known:

> The present state of the system of nature is evidently a consequence of what it was in the preceding moment, and if we conceive of an intelligence that at a given instant comprehends all the relations of the entities of this universe, it could state the respective position, motions, and general effects of all these entities at any time in the past or future.
>
> Laplace, 1814. Quoted in [48]

As a consequence, the concept of probability was interpreted as a kind of equal probability with respect to our ignorance (the subjectivist interpretation of probability).

A second attempt at defeating the ontological features of probability was made by Richard von Mises (1919). In his objective interpretation, probability was defined axiomatically in terms of more basic operational procedures for computing the relative limiting frequency of elements on a collective level (frequency interpretation). Phenomena exhibit statistical regularities measured through their mean values and dispersion. Thus, using mathematical abstractions and idealizations, the task of probability theory is to represent the connections and dependencies of observable phenomena. Probability has meaning only insofar as it can be reduced to statistics, and one then obtains the so-called statistical probability.

Kolmogorov (1933) was the first to exhibit some ontological attributes underlying the concept of probability. He used an axiomatization process based on measure theory. Having established this ontological basis, Kolmogorov began to search for conditions under which probability theory could be applied. In 1963, seeking the utility of probability in the epistemological domain, he introduced the concept of algorithmic complexity as a way of understanding how the brain processes information, by analogy with the working of computers.

Kolmogorov's great achievement was to reveal the ontological aspects of the probability concept in mathematics. Unfortunately, as we have just seen in Chap. 1, probability never appeared in physics to have any ontological features.

- The formulation of Newtonian physics and both Einstein's relativity theories slammed the door on any idea of probability. According to Einstein, such a concept has nothing to do with our ignorance. It is an important scientific concept, provided that it is placed at the epistemological level of the physical description. Thus, from the theoretical standpoint any use of probability notions should be justified on the basis of the fundamental physics of isolated systems.
- Brownian motion was reduced to the statistical physics of isolated systems.
- In statistical physics, probability enters only epistemologically, i.e., either as error due to the operational difficulties of manipulating and controlling the initial conditions of a large number of particles (e.g., in a gas), or as a problem of instabilities in the initial conditions when predicting the motion of a few isolated particles.
- In Bohm's ontological version, and in Heisenberg's quantum physics where ontology is totally absent from the quantum domain, probability enters only operationally through the statistics of acts of measurement.

Inspired by the indeterministic physics of Heisenberg for isolated systems, de Finetti developed a subjective interpretation of probability (a type of Bayesianism), where probability is a subjective degree of somebody's belief. According to this approach one arrives at the conclusion that 'probability does not exist' [39]. It is striking to note that, starting from Laplace's determinism or from Heisenberg's indeterminism, one arrives at subjective concepts of probability. This happens because of the concept of isolated systems maintained by both approaches. Isolated systems hide the ontological status of probability.

In contrast, open systems are described mathematically as stochastic processes. Hence in this chapter we present some probabilistic concepts, such as probability space, stochastic variable, probability distribution function, probability density function, characteristic function, conditional probability space, stochastic differential equations, Kolmogorov stochastic equations, and some statistical properties (e.g., mean, dispersion) which allow us to connect theory with experiment.

In our presentation of the theory of stochastic processes, we shall not be preoccupied with mathematical rigor. We only wish to point out its relevance to the concept of open systems.

2.2 Probability Space

In order to exhibit a few ontological features of probability theory, we follow the Kolmogorov approach [39,49,50]. In this context, we seek to understand the concept of open system.

Mathematically, probability is defined within the framework of set theory (see Appendix A for details of the notation). Here the main object is the probability space $(\Omega, \mathcal{B}, P)$. The space Ω is a set with an arbitrary number of elements $\omega_1, \omega_2, \ldots$, interpreted geometrically as individual points. The Borel field $\mathcal{B}$ is a class having an arbitrary number of elements $A_1, A_2, \ldots$ which are subsets of Ω, including the empty set $\emptyset$ and the whole set Ω. Geometrically, each A_k is viewed as an ensemble of points. $\mathcal{B}$ possesses an algebraic structure based on the following rules or operations involving its elements A_k:

- the union or sum of any pair A_i and A_j is in $\mathcal{B}$: $(A_i \cup A_j) \in \mathcal{B}$,
- the intersection or product of any pair A_i and A_j is in $\mathcal{B}$: $(A_i \cap A_j) \in \mathcal{B}$ or $A_i A_j \in \mathcal{B}$,
- the difference between any pair A_i and A_j is in $\mathcal{B}$: $(A_i - A_j) \in \mathcal{B}$.

Finally, the set function $P = P(A)$, the so-called probability function, quantifies each set $A \in \mathcal{B}$ by assigning a real number $P(A)$ to it:

$$P : \mathcal{B} \longrightarrow \mathbb{R} \, .$$

Its properties are (axioms of probability):

(1) $P(A) \geq 0$ is a nonnegative real scalar function.

(2) $P(\Omega) = 1$.
(3) $P(A_1 \cup A_2 \cup \ldots \cup A_n) = P(A_1) + P(A_2) + \cdots + P(A_n)$ for a finite number of countable mutually independent sets $(A_i \cap A_j) = \emptyset$, if $i \neq j$.
(3′) $P(A_1 \cup A_2 \cup \ldots) = P(A_1) + P(A_2) + \cdots$ for an infinite number of countable mutually independent sets $(A_i \cap A_j) = \emptyset$, if $i \neq j$.

According to these axioms the probability function $P(A)$ is a completely additive, nonnegative set function defined for sets $A \in \mathcal{B}$. Furthermore, to ensure that this probability function can be well defined or constructed, the probability theory must be based on the smallest class of subsets of Ω which is a Borel field, thus avoiding possible violations of its axioms.

From the above axioms, one can deduce that

$$0 \leq P(A) \leq 1 \, .$$

Consequently, we have

$$P(A) + P(A^{\mathrm{c}}) = 1 \, , \qquad P(A^{\mathrm{c}}) = 1 - P(A) \, ,$$

where $(A \cup A^{\mathrm{c}}) = \Omega$. Because $\Omega^{\mathrm{c}} = \emptyset$, we obtain $P(\emptyset) = 0$. If $A_1 \subset A_2$, then $P(A_1) \leq P(A_2)$, that is, the probability function is increasing. If the probability function of the set $A_1 \cap A_2$ is factorized, i.e.,

$$P(A_1 A_2) = P(A_1) P(A_2) \, ,$$

then A_1 and A_2 are said to be mutually independent.

In order to understand the axioms of probability theory physically, let us consider the motion of a physical system S (e.g., a Brownian particle) under the influence of an environment E. We wish to interpret the three following cases:

- What is the significance of $P = 0$?
- What is the significance of $P = 1$?
- What is the significance of $0 < P < 1$?

The theory of open systems is not a theory of everything. Considering the case $P = 0$, we are able to specify the nature of the environment E by excluding nonphysical phenomena such as metaphysical beings (e.g., gods and demons), or even the mind of the reader, from our probabilistic description.

The case $P = 1$ may be interpreted from two different standpoints. Firstly, we assume that environmental influences upon our system S occur via

- specification of initial conditions for its motion,
- specification of boundary conditions which restrict the particle motion to some closed region of the space, e.g., a Brownian particle in a box,
- all types of physical forces under consideration.

The quantification $P = 1$ of all these possibilities presupposes absolute knowledge of all interactions between the environment and the system. This is an example that illustrates the so-called Laplacian demon [51]:

> In the Introduction to his *Théorie analytiques des probabilités* Laplace envisages an all-embracing spirit possessing complete knowledge of the state of the universe at a given moment, for whom the whole universe in every detail of its existence and development would thus be completely determined. Such a spirit, knowing all forces operative in nature, and the exact positions of all the particles that make up the universe, would only have to subject these data to mathematical analysis in order to arrive at a cosmic formula that would incorporate the movements both of the largest bodies and of the lightest atoms. Nothing would be uncertain for it; future and past would lie before its gaze with the same clarity. The human mind may be seen as the copy, though weak, of such a spirit when one considers the completeness to which it has brought astronomy, but it will certainly never reach the perfection of its original. No matter how great the effort to approach it, human understanding will always remain infinitely far behind.

It should be emphasized that this way of thinking is based on non-physical or 'metaphysical' considerations. In the above, we have ascribed probability zero to the influence of interesting metaphysical entities such as gods, spirits or demons.

Does the case $P = 1$ have any physical meaning? In order to give a non-Laplacian interpretation to this question, let us suppose that our system S is isolated, i.e., there exists no environmental influence on it. Moreover, we assume that the initial and boundary conditions can be specified in a deterministic way, that is, with probability one. This interpretation reveals that the situation characterized by $P = 1$ cannot take into account novel physical properties, such as dissipative and chaotic structures (see Chap. 1) arising from the action of the environment on S. These new phenomena arise only in the case $0 < P < 1$. An isolated system follows as a very special case from an open system. In this sense the probabilistic description, where there are an infinite number of possibilities in the range $0 < P < 1$, is more complete and fundamental than the deterministic one, whose range is restricted to a unique value for P, namely, $P = 1$.

Thus, the open system notion leads us to conclude that every physical description of reality is in principle based on the concept of probability.

2.3 Stochastic Variables

2.3.1 Probabilistic Properties

Let $\Phi = \Phi(\omega)$ be a set function [49,50,52]

$$\Phi : \Omega \longmapsto \mathbb{R} ,$$

mapping each point ω of the space Ω onto points ϕ of the state space $\mathbb{R}$ (the set of real numbers). This function is said to be a stochastic variable if the algebraic structure of the Borel field is respected in $\mathbb{R}$, i.e.,

$$A = \{\omega|\Phi(\omega) \leq \phi\} \in \mathcal{B} , \qquad -\infty < \phi < \infty .$$

In a sequence of stochastic variables $\Phi_1(\omega), \Phi_2(\omega), \ldots, \Phi_n(\omega)$, defined on the same Borel field, each $\Phi_j(\omega)$, $j = 1, \ldots, n$, maps every $\omega \in \Omega$ onto a point of the n-dimensional Euclidean space $\mathbb{R}^n$.

Physical quantities, such as position and velocity are examples of stochastic variables defined on the whole real line, while energy is an example of stochastic function restricted to the interval $(0, \infty)$.

Distribution Function

A stochastic variable $\Phi(\omega)$ has probabilistic meaning if it is defined within the framework of a probability space $(\Omega, \mathcal{B}, P)$. We need therefore to quantify numerically, according to the prescription $0 \leq P \leq 1$, each possible value ϕ of the variable $\Phi(\omega)$ in the state space $\mathbb{R}$ (our space Ω). Here the probability function $P(A)$ turns out to be

$$P(\{\Phi(\omega) \leq \phi\}) = F_\Phi(\phi)$$

and is called the distribution function of $\Phi(\omega)$.

For n stochastic variables, the function

$$F_{\Phi_1\Phi_2\ldots\Phi_n}(\phi_1,\phi_2,\ldots,\phi_n) = P(\{(\Phi_1 \le \phi_1)\cap(\Phi_2 \le \phi_2)\cap\ldots\cap(\Phi_n \le \phi_n)\})$$

is called the joint distribution function. Consider the set of s variables $\{\Phi_1,\Phi_2,\ldots,\Phi_s\}$, with $s < n$. The joint probability distribution function

$$F_{\Phi_1\Phi_2\ldots\Phi_n}(\phi_1,\phi_2,\ldots,\phi_s,\infty,\ldots,\infty) = F_{\Phi_1\Phi_2\ldots\Phi_s}(\phi_1,\phi_2,\ldots,\phi_s)$$

is referred to as the joint marginal distribution function. The variables $\Phi_1,\Phi_2,\ldots,\Phi_n$ are mutually independent if

$$F_{\Phi_1\Phi_2\ldots\Phi_n}(\phi_1,\phi_2,\ldots,\phi_n) = F_{\Phi_1}(\phi_1)F_{\Phi_2}(\phi_2)\ldots F_{\Phi_n}(\phi_n)\,.$$

Density Function

As we are supposing that $F_\Phi(\phi)$ is continuous and differentiable almost everywhere, there exists a derivative

$$D_\Phi(\phi) = \frac{\mathrm{d}}{\mathrm{d}\phi}F_\Phi(\phi) \qquad \text{or} \qquad \mathrm{d}F_\Phi(\phi) = D_\Phi(\phi)\mathrm{d}\phi\,,$$

called the density function of $\Phi(\omega)$. It has the following properties:

- $D_\Phi(\phi) \ge 0\,,$
- $\int_{-\infty}^{\phi} D_\Phi(y)\mathrm{d}y = F_\Phi(\phi)\,,$
- $\int_{-\infty}^{\infty} D_\Phi(\phi)\mathrm{d}\phi = 1\,.$

The joint density function for n stochastic variables is

$$D_{\Phi_1\Phi_2\ldots\Phi_n}(\phi_1,\phi_2,\ldots,\phi_n) = \frac{\partial^n}{\partial\phi_1\partial\phi_2\ldots\partial\phi_n}F_{\Phi_1\Phi_2\ldots\Phi_n}(\phi_1,\phi_2,\ldots,\phi_n)\,,$$

provided that the partial derivatives exist.

In terms of the joint density function the stochastic variables $\Phi_1,\ldots,\Phi_n$ are said to be mutually independent if $D_{\Phi_1\ldots\Phi_n}(\phi_1,\ldots,\phi_n)$ is factorized

$$D_{\Phi_1\ldots\Phi_n}(\phi_1,\ldots,\phi_n) = D_{\Phi_1}(\phi_1)\ldots D_{\Phi_n}(\phi_n)\,.$$

Characteristic Function

The characteristic function of a stochastic variable $\Phi = \Phi(\omega)$ is given by the Fourier transform of the density function $D_\Phi(\phi)$:

$$f_\Phi(u) = \int_{-\infty}^{\infty} \mathrm{e}^{\mathrm{i}u\phi} D_\Phi(\phi)\mathrm{d}\phi \,,$$

where u is arbitrary and real-valued. Even though $D_\Phi(\phi) \geq 0$, the characteristic function $f_\Phi(u)$ may have nonpositive values. The inverse Fourier transform yields

$$D_\Phi(\phi) = \frac{1}{2\pi} \int_{-\infty}^{\infty} \mathrm{e}^{-\mathrm{i}u\phi} f_\Phi(u)\mathrm{d}u \,.$$

The generalization of the characteristic function to more than two variables is

$$\begin{aligned} &f_{\Phi_1 \ldots \Phi_n}(u_1, \ldots, u_n) \\ &= \int_{-\infty}^{\infty} \cdots \int_{-\infty}^{\infty} \mathrm{e}^{\mathrm{i}(u_1\phi_1 + \cdots + u_n\phi_n)} F_{\Phi_1 \ldots \Phi_n}(\phi_1, \ldots, \phi_n)\mathrm{d}\phi_1 \ldots \mathrm{d}\phi_n \,, \end{aligned}$$

where $u_1, \ldots, u_n$ are arbitrary real variables, while its inverse transform is

$$\begin{aligned} &F_{\Phi_1 \ldots \Phi_n}(\phi_1, \ldots, \phi_n) \\ &= \frac{1}{(2\pi)^n} \int_{-\infty}^{\infty} \cdots \int_{-\infty}^{\infty} \mathrm{e}^{-\mathrm{i}(u_1\phi_1 + \cdots + u_n\phi_n)} f_{\Phi_1 \ldots \Phi_n}(u_1, \ldots, u_n)\mathrm{d}u_1 \ldots \mathrm{d}u_n \,. \end{aligned}$$

In summary, in an abstract way, the probabilistic nature of a stochastic variable $\Phi(\omega)$ is revealed by specifying:

- the set of its possible values ϕ on the space $\Omega = \mathbb{R} \in \mathcal{B}$,
- its distribution function $F_\Phi(\phi)$, its density function $D_\Phi(\phi)$, and its characteristic function $f_\Phi(u)$.

2.3.2 Statistical Properties

Physical quantities (e.g., the position and momentum of a Brownian particle) are represented by stochastic variables whose probabilistic content is studied by means of the abstract concepts of distribution function, density function and characteristic function. The theory of probability also provides the statistical properties of the stochastic variables through quantities such as the mean value and dispersion which can in turn be connected to empirical observations [52,53].

Moments

For a stochastic variable $\Phi(\omega)$ with density function $D_\Phi(\phi)$, we assume that the moments defined by

$$\alpha_n \equiv \langle \Phi^n \rangle = \int_{-\infty}^{\infty} \phi^n D_\Phi(\phi) \mathrm{d}\phi \qquad (n = 1, 2, 3, \ldots)$$

are finite. The particular case $\alpha_1 = \langle \Phi \rangle$ is the expected value or mean of Φ. The characteristic function of Φ is related to the moments α_n through the formula

$$f_\Phi(u) = \langle \mathrm{e}^{\mathrm{i}u\Phi} \rangle = 1 + \sum_{k=1}^{\infty} \frac{(\mathrm{i}u)^k}{k!} \alpha_k \ .$$

This equation means that all the statistical properties of a stochastic variable Φ are determined by the infinite set of moments α_n.

The so-called central moments are moments of Φ taken with respect to its mean $\alpha_1 = \langle \Phi \rangle$, viz.,

$$\mu_n \equiv \langle (\Phi - \alpha_1)^n \rangle = \int_{-\infty}^{\infty} (\phi - \alpha_1)^n D_\Phi(\phi) \mathrm{d}\phi \ .$$

We note that $\mu_1 = 0$, while $\mu_2 = \langle (\Phi - \alpha_1)^2 \rangle$ measures the dispersion, variance, or fluctuation of Φ about its mean. The standard deviation is $\sigma_\Phi = \sqrt{\mu_2}$.

As an example, let us consider a stochastic variable $\Phi(\omega)$ characterized by the density function $D_\Phi(\phi) = \delta(\phi - \varphi)$, where the Dirac delta function is defined by its properties under the integral sign:

$$\int_{-\infty}^{\infty} g(y)\delta(y - x)\mathrm{d}y = g(x) \ ,$$

for any continuous function $g(y)$. The expected value of this Φ is then

$$\langle \Phi \rangle = \int_{-\infty}^{\infty} \phi\delta(\phi - \varphi)\mathrm{d}\phi = \varphi \ .$$

Its characteristic function has the form

$$f_\Phi(u) = \langle \mathrm{e}^{\mathrm{i}u\Phi} \rangle = \mathrm{e}^{\mathrm{i}u\varphi} \ .$$

This type of stochastic variable is characterized by the mean $\langle \Phi \rangle = \varphi$, while all its central moments are zero.

Stochastic variables Φ with $D_\Phi(\phi) = \delta(\phi - \varphi)$ as density functions are called deterministic variables. The probability distribution function associated with a deterministic variable is

$$P(\{\Phi \le \phi\}) = F_\Phi(\phi) = \int_{\phi-\epsilon}^{\phi+\epsilon} \delta(\phi - \varphi) \mathrm{d}\phi = 1 ,$$

for all $\epsilon > 0$.

Joint Moments

The joint moments of Φ and Ψ are

$$\alpha_{nm} \equiv \langle \Phi^n \Psi^m \rangle = \int_{-\infty}^{\infty} \int_{-\infty}^{\infty} \phi^n \psi^m D_{\Phi\Psi}(\phi, \psi) \mathrm{d}\phi \mathrm{d}\psi ,$$

when they exist. Special cases are $\alpha_{0m} = \langle \Psi^m \rangle$ and $\alpha_{n0} = \langle \Phi^n \rangle$.

The joint central moments of Φ and Ψ are defined by

$$\begin{aligned} \mu_{nm} &\equiv \Big\langle (\Phi - \langle \Phi \rangle)^n (\Psi - \langle \Psi \rangle)^m \Big\rangle \\ &= \int_{-\infty}^{\infty} \int_{-\infty}^{\infty} (\phi - \langle \Phi \rangle)^n (\psi - \langle \Psi \rangle)^m D_{\Phi\Psi}(\phi, \psi) \mathrm{d}\phi \mathrm{d}\psi . \end{aligned}$$

The quantities μ_{20} and μ_{02} are the respective variances of Φ and Ψ:

$$\mu_{20} = \Big\langle (\Phi - \langle \Phi \rangle)^2 \Big\rangle = \sigma_\Phi^2 , \qquad \mu_{02} = \Big\langle (\Psi - \langle \Psi \rangle)^2 \Big\rangle = \sigma_\Psi^2 ,$$

while μ_{11} measures the covariance of Φ and Ψ:

$$\mu_{11} = \Big\langle (\Phi - \langle \Phi \rangle)(\Psi - \langle \Psi \rangle) \Big\rangle = \langle \Phi\Psi \rangle - \langle \Phi \rangle \langle \Psi \rangle = \sigma_{\Phi\Psi} .$$

The normalized covariance is the dimensionless quantity

$$\rho_{\Phi\Psi} = \frac{\mu_{11}}{\sqrt{\mu_{20}\mu_{02}}} = \frac{\sigma_{\Phi\Psi}}{\sqrt{\sigma_\Phi \sigma_\Psi}} , \qquad |\rho_{\Phi\Psi}| \le 1 ,$$

and is called the correlation coefficient of Φ and Ψ.

The stochastic variables Φ and Ψ are said to be uncorrelated if $\rho_{\Phi\Psi} = 0$. This implies that $\mu_{11} = 0$ or $\langle \Phi\Psi \rangle = \langle \Phi \rangle \langle \Psi \rangle$. In the case $\langle \Phi\Psi \rangle = 0$, Φ and Ψ are orthogonal.

If Φ and Ψ are independent, i.e., if

$$D_{\Phi\Psi}(\phi, \psi) = D_\Psi(\phi) D_\Phi(\phi) ,$$

then they are uncorrelated: $\langle \Phi\Psi \rangle = \langle \Phi \rangle \langle \Psi \rangle$ or $\rho_{\Phi\Psi} = 0$. A lack of correlation between Φ and Ψ does not necessarily imply that they are independent.

Using the generalization of the characteristic function to n stochastic variables,

$$f(u_1, \ldots, u_n) = \left\langle e^{i(u_1 \Phi_1 + \cdots + u_n \Phi_n)} \right\rangle ,$$

it is also possible to find generalized expressions involving moments [53].

The deterministic character of a stochastic variable can be extended to a sequence of n variables. For instance, there is no correlation between Φ_1 and Φ_2, since $D_{\Phi_1 \Phi_2} = \delta(\phi_1 - \varphi_1)\delta(\phi_2 - \varphi_2)$. The joint characteristic function is simply

$$f(u_1, \ldots, u_n) = \left\langle e^{i(u_1 \Phi_1 + \cdots + u_n \Phi_n)} \right\rangle = e^{i(u_1 \varphi_1 + \cdots + u_n \varphi_n)} .$$

2.4 Conditional Probability Space

Another way of quantifying the elements of the Borel field $\mathcal{B}$ is by means of the conditional probability function defined as [49,50]

$$W(A|B) = \frac{P(AB)}{P(B)} \qquad [P(B) > 0] ,$$

which is indeed a set function $W(A)$. The notation $W(A|B)$ only indicates in an explicit way that the probability of A is conditioned by the probability of B. Its properties are:

- If $B \subset A$, then $W(A|B) = 1$. From this it follows that $W(A|\Omega) = 1$ and $W(A|A) = 1$.
- If $A \subset B$ then

$$W(A|B) = \frac{P(A)}{P(B)} \geq P(A) .$$

In terms of the conditional probability function W, the axioms of probability theory read:

(1) $W(A|B) \geq 0$ is a real scalar function satisfying the Borel algebraic structure.
(2) $W(\Omega|B) = 1$.
(3) $W(A_1 \cup A_2 \cup \ldots A_n|B) = W(A_1|B) + W(A_2|B) + \cdots + W(A_n|B)$, if $A_1, \ldots, A_n$ are mutually independent.

There are some theorems relating W and P. If we consider an arbitrary set X and $A_1 + A_2 + \cdots + A_n = \Omega$, then it follows that

$$P(X) = W(X|A_1)P(A_1) + W(X|A_2)P(A_2) + \cdots + W(X|A_n)P(A_n) .$$

Bayes' theorem is given by

$$W(A_i|X) = \frac{P(A_i)W(X|A_i)}{\sum_{j=1}^{n} P(A_i)W(X|A_i)} \quad (i = 1, 2, 3, \ldots, n) .$$

In terms of the conditional probability function, the independence of A_1 and A_2 is expressed by

$$W(A_1|A_2) = \frac{P(A_1)P(A_2)}{P(A_2)} = P(A_1) ,$$

and the probability space can be written as $(\Omega, \mathcal{B}, W)$.

2.4.1 Conditional Distribution Function

Let $\Phi(\omega)$ and $\Psi(\omega)$ be two stochastic variables on the space Ω. From the definition of the conditional probability function, we have

$$W(\{\Phi(\omega) \leq \phi | \Psi(\omega) \leq \psi\}) = \frac{P(\{\Phi(\omega) \leq \phi\} \cap \{\Psi(\omega) \leq \psi\})}{P(\{\Psi(\omega) \leq \psi\})} .$$

The function

$$F_{\Phi|\Psi}(\phi|\psi) \equiv W(\{\Phi \leq \phi | \Psi(\omega) \leq \psi\})$$

is called the conditional distribution function of Φ given Ψ. It has the same properties as $F_{\Phi}(\phi)$:

- $F(\infty|\psi) = 1$,
- $F(-\infty|\psi) = 0$.

2.4.2 Conditional Density Function

The conditional density function of Φ is

$$D_{\Phi|\Psi}(\phi|\psi) = \frac{\mathrm{d}F_{\Phi|\Psi}(\phi|\psi)}{\mathrm{d}\phi} = \frac{D_{\Phi\Psi}(\phi, \psi)}{D_{\Psi}(\psi)} .$$

Its generalization to more than two stochastic variables is

$$D_{\Phi_1\ldots\Phi_n|\Psi_1\ldots\Psi_m}(\phi_1,\ldots,\phi_n|\psi_1,\ldots,\psi_m) = \frac{D_{\Phi_1\ldots\Phi_n\Psi_1\ldots\Psi_m}(\phi_1,\ldots,\phi_n,\psi_1,\ldots,\psi_m)}{D_{\Psi_1\ldots\Psi_m}(\psi_1,\ldots,\psi_m)} ,$$

which satisfies the normalization property

$$\int_{-\infty}^{\infty}\cdots\int_{-\infty}^{\infty} D_{\Phi_1\ldots\Phi_n|\Psi_1\ldots\Psi_m}(\phi_1,\ldots,\phi_n|\psi_1,\ldots,\psi_m)\mathrm{d}\phi_1\ldots\mathrm{d}\phi_n = 1 .$$

All statistical properties can be expressed in terms of the conditional probability density functions. For instance, the mean of Φ conditioned by Ψ is

$$\langle(\Phi|\Psi)\rangle = \int_{-\infty}^{\infty} \phi D_{\Phi|\Psi}(\phi|\Psi)\mathrm{d}\phi ,$$

which is also a stochastic variable.

2.5 Stochastic Processes

Up to now, probabilistic concepts have been defined within a static framework, i.e., independently of their behavior with respect to any parameter. The probability space alone is thus insufficient to describe physical phenomena, since these vary in time, for example.

Bringing time into probability theory generates the concept of a stochastic process [54]. Its definition is only possible within the framework of the Kolmogorov space $\mathcal{K} = (\Omega, \mathcal{B}, P, \boldsymbol{T})$, where $\boldsymbol{T}$ is a set whose elements t are time parameters, assumed to be continuous and with range $0 \leq t < \infty$ in the set of real numbers. Then a physical system is considered as a stochastic process if its probabilistic properties change in time.

On the basis of the Kolmogorov space $\mathcal{K}$, we could then introduce a point function $\Phi = \Phi(\omega, t)$,

$$\Phi : \Omega \times \boldsymbol{T} \longmapsto \mathbb{R} ,$$

taking all $\omega \in \Omega$ and $t \in \boldsymbol{T}$ to a point of the state space $\mathbb{R}$ (the set of real numbers).

If the stochastic function $\Phi(\omega, t)$ is to have some probabilistic content, we have to associate with it the probability function

$$P(\{\Phi(\omega,t) \leq \phi(t)\}) = F_{\Phi}(\phi, t) ,$$

called the distribution function of the stochastic process $\Phi(\omega, t)$. Its derivative leads to the density function

$$\frac{\mathrm{d}}{\mathrm{d}t} F_\Phi(\phi, t) = D_\Phi(\phi, t),$$

while the Fourier transform of $D_\Phi(\phi, t)$ yields the characteristic function

$$f_\Phi(u, t) = \int_{-\infty}^{\infty} \mathrm{e}^{\mathrm{i}u\phi} D_\Phi(\phi, t) \mathrm{d}\phi ,$$

where $u = u(t)$ is an arbitrary, real-valued function.

We now assume that the time evolution of $\Phi(\omega, t)$ is governed by the ordinary differential equation [55–58]

$$\frac{\mathrm{d}}{\mathrm{d}t} \Phi(\omega, t) = \Xi(\Phi, \omega, t) . \tag{2.1}$$

Because its solution is a function defined on the Kolmogorov space $\mathcal{K} = (\Omega, \mathcal{B}, P, \boldsymbol{T})$, we call this a stochastic differential equation. On averaging (2.1) according to

$$\langle \cdots \rangle = \int_{-\infty}^{\infty} \cdots D_\Phi(\phi, t) \mathrm{d}\phi , \tag{2.2}$$

we have

$$\frac{\mathrm{d}}{\mathrm{d}t} \langle \Phi(t) \rangle = \langle \Xi(\Phi, t) \rangle . \tag{2.3}$$

If we use $D_\Phi(\phi, t) = \delta(\phi - \varphi)$ in (2.3), we obtain the deterministic differential equation

$$\frac{\mathrm{d}}{\mathrm{d}t} \varphi(t) = \xi(\varphi, t) . \tag{2.4}$$

Equations (2.1) and (2.4) are solved once the initial conditions and the boundary conditions have been specified. These may be either of a stochastic or a deterministic nature. Thus we may have an equation of type (2.1) with deterministic initial conditions or an equation of type (2.4) with stochastic initial conditions. Mathematical properties of the solutions, such as existence, uniqueness, differentiability and continuity, are investigated taking into account the fact that a certain probability in the range $0 \leq P < 1$ is associated with each solution of (2.1), while every solution of (2.4) is quantified with probability 1.

The dynamics of a set of n stochastic processes $\Phi_k(\omega, t)$, $k = 1, 2, \ldots, n$, is determined by the system of ordinary differential equations

$$\begin{aligned} \dot{\Phi}_1 &\equiv \frac{\mathrm{d}}{\mathrm{d}t} \Phi_1 = K_1(\Phi_1, \ldots, \Phi_n, \omega, t) , \\ &\vdots \\ \dot{\Phi}_n &\equiv \frac{\mathrm{d}}{\mathrm{d}t} \Phi_n = K_n(\Phi_1, \ldots, \Phi_n, \omega, t) . \end{aligned} \tag{2.5}$$

Geometrically, this system of stochastic differential equations (2.5), or simply this stochastic system, may be studied using the concept of phase space, denoted by Γ. This is a space with n dimensions and axes $\Phi_1, \ldots, \Phi_n$. The state of the stochastic system is therefore represented by a finite region $\mathcal{D} \subset \Gamma$, while the time evolution of each point in $\mathcal{D}$ follows a trajectory or orbit with certain stochastic properties, such as a non-Euclidean structure (e.g., the fractal geometry underlying the Brownian trajectories [24]). All the points or an ensemble of points in $\mathcal{D}$ generate a motion of the region $\mathcal{D}$. Using a hydrodynamical analogy, we can derive the motion of this region as determined by the stochastic Liouville equation [55]:

$$\frac{\partial \mathcal{P}}{\partial t} + \sum_{i=1}^{n} K_i \frac{\partial \mathcal{P}}{\partial \Phi_i} = -\mathcal{P} \sum_{i=1}^{n} \frac{\partial K_i}{\partial \Phi_i} , \tag{2.6}$$

which can be solved once (stochastic or deterministic) initial and boundary conditions have been imposed. The stochastic function $\mathcal{P}(\Phi_1, \ldots, \Phi_n, t)$ stands for the density of points in $\mathcal{D}$, i.e.,

$$\frac{\mathrm{d}}{\mathrm{d}\mathcal{V}} \mathcal{N} = \mathcal{P} , \tag{2.7}$$

where

$$N = \int_{\mathcal{D}} \mathrm{d}\mathcal{N} = \int_{\mathcal{D}} \mathcal{P} \mathrm{d}\Phi_1 \ldots \mathrm{d}\Phi_n \tag{2.8}$$

determines the number of points within the region $\mathcal{D}$ of phase space. (Notice that this N is also a stochastic function.)

Using the joint density function

$$D_{\Phi_1 \ldots \Phi_n}(\phi_1, \ldots \phi_n, t) = \delta(\phi_1 - \varphi_1) \ldots \delta(\phi_n - \varphi_n) , \tag{2.9}$$

we average the system of stochastic differential equations (2.5), reducing it to the deterministic dynamical system [3]

$$\begin{aligned} \dot{\varphi}_1 \equiv \frac{\mathrm{d}}{\mathrm{d}t} \varphi_1 &= K_1(\varphi_1, \ldots, \varphi_n, t) , \\ &\vdots \\ \dot{\varphi}_n \equiv \frac{\mathrm{d}}{\mathrm{d}t} \varphi_n &= K_n(\varphi_1, \ldots, \varphi_n, t) , \end{aligned} \tag{2.10}$$

whose solutions may be obtained from (deterministic or stochastic) initial and boundary conditions.

On the other hand, by averaging the stochastic Liouville equation (2.6) with (2.9), we arrive at the deterministic Liouville equation

$$\frac{\partial D_{\mathcal{P}}}{\partial t} + \sum_{i=1}^{n} K_i \frac{\partial D_{\mathcal{P}}}{\partial \varphi_i} = -D_{\mathcal{P}} \sum_{i=1}^{n} \frac{\partial K_i}{\partial \varphi_i} , \tag{2.11}$$

where $D_{\mathcal{P}} \equiv D_{\mathcal{P}}(\varphi_1, \ldots, \varphi_n, t)$ should be interpreted as a probability density function. Alternatively (2.11) can be generated from (2.10).

The probabilistic properties in time of the stochastic process $\Phi(\omega, t)$ are investigated by means of an equation of motion for the probability density function $D_\Phi(\phi, t)$. To find this equation, we consider the following transformation $(\phi', t) \mapsto (\phi, t')$:

$$\phi = \phi' + \Delta\phi , \qquad t' = t + \Delta t , \tag{2.12}$$

where $\Delta\phi$ and Δt are (finite) increments in ϕ and t, respectively. We now use the concept of conditional probability density, defined by

$$W(\phi, t'|\phi', t) = \frac{D_\Phi(\phi, t'; \phi', t)}{D_\Phi(\phi', t)} , \tag{2.13}$$

from which we derive the Einstein identity [36]

$$D_\Phi(\phi, t') = \int_{-\infty}^{\infty} W(\phi, t'|\phi', t) D_\Phi(\phi', t) \mathrm{d}\phi' . \tag{2.14}$$

The characteristic function for the increment $\Delta\Phi = \Phi(t + \Delta t) - \Phi(t)$ is expressed in terms of the conditional density W as follows [53]:

$$\langle \mathrm{e}^{\mathrm{i}u\Delta\Phi} \rangle = \int_{-\infty}^{\infty} \mathrm{e}^{\mathrm{i}u\Delta\phi} W(\phi, t'|\phi', t) \mathrm{d}\phi' , \tag{2.15}$$

which has inverse

$$W(\phi, t'|\phi', t) = \frac{1}{2\pi} \int_{-\infty}^{\infty} \mathrm{e}^{-\mathrm{i}u\Delta\phi} \langle \mathrm{e}^{\mathrm{i}u\Delta\Phi} \rangle \mathrm{d}u . \tag{2.16}$$

Now making use of the expansion

$$\langle \mathrm{e}^{\mathrm{i}u\Delta\Phi} \rangle = \sum_{s=0}^{\infty} \frac{(\mathrm{i}u)^s}{s!} \langle (\Delta\Phi)^s \rangle , \tag{2.17}$$

equation (2.16) becomes

$$W(\phi, t'|\phi', t) = \sum_{s=0}^{\infty} \frac{(-1)^s}{s!} \langle (\Delta\Phi)^s \rangle \frac{\partial^s}{\partial \phi^s} \delta(\phi - \phi') . \tag{2.18}$$

Inserting (2.18) into (2.14) and dividing the resulting equation by Δt, we arrive at

$$\frac{D_{\Phi}(\phi, t+\Delta t) - D_{\Phi}(\phi, t)}{\Delta t} = \sum_{s=1}^{\infty} \frac{(-1)^s}{s!} \frac{\partial^s}{\partial \phi^s} \left[\frac{\langle (\Delta \Phi)^s \rangle}{\Delta t} D_{\Phi}(\phi, t) \right] . \tag{2.19}$$

Mathematically, we can consider the increment Δt as an infinitesimal, so that the limit $\Delta t \to 0$ is permissible. This procedure leads to the partial differential equation

$$\frac{\partial D_{\Phi}(\phi, t)}{\partial t} = \sum_{s=1}^{\infty} \frac{(-1)^s}{s!} \frac{\partial^s}{\partial \phi^s} \left[B^{(s)}(\phi, t) D_{\Phi}(\phi, t) \right] , \tag{2.20}$$

where the coefficients

$$B^{(s)}(\phi, t) = \lim_{\Delta t \to 0} \frac{\langle (\Delta \Phi)^s \rangle}{\Delta t} \tag{2.21}$$

are calculated from the stochastic differential equation (2.1):

$$\langle (\Delta \Phi)^s \rangle \equiv \langle [\Phi(t+\Delta t) - \Phi(t)]^s \rangle = \int_t^{t+\Delta t} \left\langle [\Xi(\Phi, \omega, t')]^s \right\rangle \mathrm{d}t' , \tag{2.22}$$

with

$$\int_t^{t+\Delta t} \langle [\Xi(\Phi, \omega, t')]^s \rangle \mathrm{d}t' = \int_{-\infty}^{\infty} \int_t^{t+\Delta t} [\Xi(\phi, t')]^s D_{\phi}(\phi, t') \mathrm{d}t' \mathrm{d}\phi . \tag{2.23}$$

Because the solution of (2.20) is defined on the Kolmogorov space $\mathcal{K} = (\Omega, \mathcal{B}, P, \boldsymbol{T})$, we call it the Kolmogorov stochastic equation. [This equation can be extended to n stochastic processes $\Phi_1(t), \ldots, \Phi_n(t)$.]

Following an analogous procedure to the one used to derive (2.20), we can derive the so-called backwards Kolmogorov stochastic equation [53]:

$$\frac{\partial}{\partial t} D_{\Phi}(\phi', t) = -\sum_{s=1}^{\infty} \frac{1}{s!} B^{(s)}(\phi', t) \frac{\partial^s}{\partial \phi'^s} D_{\Phi}(\phi', t) ,$$

which is the adjoint of (2.20).

According to the Pawula theorem [59], if the coefficients $B^{(s)}$ of (2.20) are finite for every s and if $B^{(s)} = 0$ for some even s, then $B^{(s)} = 0$ for all $s \geq 3$. As a consequence of this theorem, the Kolmogorov stochastic equation (2.20) does not violate the positivity of $D_{\Phi}(\phi, t)$ in at least three cases:

- For $s = 1$, we have

$$\frac{\partial D_{\Phi}(\phi,t)}{\partial t} = -\frac{\partial}{\partial \phi}\left[B^{(1)}(\phi,t)D_{\Phi}(\phi,t)\right] . \tag{2.24}$$

  The deterministic Liouville equation (2.11) is an example of this case.
- For $s = 2$, we have the one-variable Fokker–Planck equation (see Chap. 3):

$$\frac{\partial D_{\Phi}(\phi,t)}{\partial t} = -\frac{\partial}{\partial \phi}\left[B^{(1)}(\phi,t)D_{\Phi}(\phi,t)\right] + \frac{\partial^2}{\partial \phi^2}\left[B^{(2)}(\phi,t)D_{\Phi}(\phi,t)\right] , \tag{2.25}$$

  with coefficients

$$B^{(1)}(\phi,t) = \lim_{\Delta t \to 0} \frac{\langle(\Delta\Phi)\rangle}{\Delta t} , \tag{2.26}$$

$$B^{(2)}(\phi,t) = \lim_{\Delta t \to 0} \frac{\langle(\Delta\Phi)^2\rangle}{\Delta t} . \tag{2.27}$$

- For $s \to \infty$, we have the proper equation of motion (2.20).

It should be noted that these three conditions are only necessary conditions for the compatibility of the Kolmogorov stochastic equation (2.20) with the axioms of probability theory. In general, additional considerations are required concerning the limiting process in (2.21) to obtain the coefficients $B^{(s)}(\phi,t)$.

In summary, a complete investigation of the probabilistic and statistical properties of stochastic processes requires a study of both equations of motion:

- the stochastic differential equation (2.1) for $\Phi(\omega,t)$,
- the Kolmogorov stochastic equation (2.20) for $D_{\Phi}(\phi,t)$.

In the above, we have presented some general properties of stochastic processes. We now turn to the study of some specific properties. To this end, we visualize the stochastic function $\Phi(\omega,t)$ in two complementary representations: the fixed-time representation and the fixed-point representation.

2.5.1 Fixed-Time Representation of Stochastic Processes

Probabilistic Properties

The fixed-time representation is characterized by regarding the stochastic process $\Phi = \Phi(\omega, t)$ as indexed by arbitrary fixed values of the parameter t, i.e.,

$$\Phi(\omega, t_i) = \int_{-\infty}^{\infty} \Phi(\omega, t)\delta(t - t_i)\mathrm{d}t .$$

Thus for each fixed value t_i, $i = 1, 2, \ldots, n$, there corresponds a stochastic variable. A stochastic process is seen as a collection or family of $\Phi(\omega, t_i)$:

$$\Phi(\omega, t) = [\Phi(\omega, t_1), \Phi(\omega, t_2), \ldots, \Phi(\omega, t_n)] .$$

The fixed-time representation of a stochastic process is convenient for practical purposes, for example, for a measurement device in which time can be considered as a discrete quantity [56].

Associated with such a family of stochastic variables, there exists an arbitrary number of distribution functions:

$$\begin{aligned} &F_{\Phi}^{(n)}(\phi_1, t_1; \phi_2, t_2; \ldots; \phi_n, t_n) \\ &\qquad = P(\{\Phi(\omega, t_1) \le \phi(t_1), \ldots, \Phi(\omega, t_n) \le \phi(t_n)\}) , \end{aligned}$$

and the corresponding joint density functions

$$\begin{aligned} D_{\Phi}^{(1)}(\phi_1, t_1) &= \frac{\partial}{\partial \phi_1} F_{\Phi}^{(1)}(\phi_1, t_1) , \\ D_{\Phi}^{(2)}(\phi_1, t_1; \phi_2, t_2) &= \frac{\partial^2}{\partial \phi_1 \partial \phi_2} F_{\Phi}^{(2)}(\phi_1, t_1; \phi_2, t_2) , \\ &\vdots \\ D_{\Phi}^{(n)}(\phi_1, t_1; \phi_2, t_2; \ldots; \phi_n, t_n) &= \frac{\partial^n}{\partial \phi_1 \ldots \partial \phi_n} F_{\Phi}^{(n)}(\phi_1, t_1; \ldots; \phi_n, t_n) . \end{aligned}$$

Hence, in order to characterize a stochastic process completely, we must specify an arbitrary number of n joint probability density functions. In the limiting situation $n \to \infty$, one introduces the notion of probability functional [53,55].

The density functions $D_{\Phi}^{(n)}$ of the stochastic process $\Phi(\omega, t)$ obey the Kolmogorov consistency conditions:

- symmetry condition: swapping any $\phi(t_k)$ and $\phi(t_l)$ leaves $D_\Phi^{(n)}$ invariant:

$$D_\Phi^{(n)}(\phi_1, t_1; \ldots; \phi_k, t_k; \phi_l, t_l; \ldots; \phi_n, t_n) \qquad (2.28)$$
$$= D_\Phi^{(n)}(\phi_1, t_1; \ldots; \phi_l, t_l; \phi_k, t_k; \ldots; \phi_n, t_n) ,$$

- compatibility condition: given the n th order joint density function $D_\Phi^{(n)}$, we obtain all the $D_\Phi^{(r)}$ with $r < n$:

$$\int_{-\infty}^{\infty} \cdots \int_{-\infty}^{\infty} D_\Phi^{(n)}(\phi_1, t_1; \ldots; \phi_n, t_n) \mathrm{d}\phi_{r+1} \ldots \mathrm{d}\phi_n \qquad (2.29)$$
$$= D_\Phi^{(n-r)}(\phi_1, t_1; \ldots; \phi_r, t_r) .$$

The hierarchy of probability density functions

$$\begin{gathered} D_\Phi^{(1)}(\phi_1, t_1) , \\ D_\Phi^{(2)}(\phi_1, t_1; \phi_2, t_2) , \\ \vdots \\ D_\Phi^{(n)}(\phi_1, t_1; \ldots; \phi_n, t_n) , \end{gathered}$$

is also given in terms of the conditional density functions by

$$W^{(n-k|k)}(\phi_n, t_n; \ldots; \phi_{k+1}, t_{k+1} | \phi_1, t_1; \ldots; \phi_k, t_k)$$
$$= \frac{D_\Phi^{(n)}(\phi_1, t_1; \ldots; \phi_k, t_k; \phi_{k+1}, t_{k+1}; \ldots; \phi_n, t_n)}{D_\Phi^{(k)}(\phi_1, t_1; \ldots; \phi_k, t_k)} .$$

Let us consider the case $k = 1$, i.e.,

$$D_\Phi^{(2)}(\phi_1, t_1; \phi_2, t_2) = W^{(1|1)}(\phi_2, t_2 | \phi_1, t_1) D_\Phi^{(1)}(\phi_1, t_1) ,$$
$$D_\Phi^{(3)}(\phi_1, t_1; \phi_2, t_2; \phi_3, t_3) = W^{(2|1)}(\phi_3, t_3; \phi_2, t_2; | \phi_1, t_1) D_\Phi^{(1)}(\phi_1, t_1) ,$$
$$\vdots$$
$$D_\Phi^{(n)}(\phi_1, t_1; \ldots; \phi_n, t_n)$$
$$= W^{(n-1|1)}(\phi_n, t_n; \ldots; \phi_{n-2}, t_{n-2} | \phi_1, t_1) D_\Phi^{(1)}(\phi_1, t_1) .$$

Using the compatibility condition (2.29), we obtain

$$D_\Phi^{(1)}(\phi_n, t_n) = \int_{-\infty}^{\infty} \cdots \int_{-\infty}^{\infty} W^{(n-1|1)}(\phi_n, t_n; \ldots; \phi_2, t_2 | \phi_1, t_1) \qquad (2.30)$$
$$\times D_\Phi^{(1)}(\phi_1, t_1) \mathrm{d}\phi_1 \ldots \mathrm{d}\phi_{n-1} .$$

The conditional density $W^{(n-1|1)}$ is interpreted as a transition density function or propagator because it changes $D_{\Phi}^{(1)}$ from t_1 to t_n, taking into account the intermediate steps. As $n \to \infty$, (2.30) can be interpreted within the framework of functional calculus, i.e., functional derivatives, functional integrals, and other properties [60].

Another way of specifying the probabilistic nature of a stochastic process is by means of a sequence of characteristic functions:

$$\begin{aligned} f_{\Phi}^{(1)}(u_1, t_1) &= \left\langle \mathrm{e}^{\mathrm{i}u_1\Phi(t_1)} \right\rangle \ , \\ f_{\Phi}^{(2)}(u_1, t_1; u_2, t_2) &= \left\langle \mathrm{e}^{\mathrm{i}[u_1\Phi(t_1)+u_2\Phi(t_2)]} \right\rangle \ , \\ &\vdots \\ f_{\Phi}^{(n)}(u_1, t_1; \ldots; u_n, t_n) &= \left\langle \mathrm{e}^{\mathrm{i}[u_1\Phi(t_1)+\cdots+u_n\Phi(t_n)]} \right\rangle \ . \end{aligned}$$

In the limiting case $n \to \infty$, we obtain the complete characterization of $\Phi(\omega, t)$ using the characteristic functional

$$f[u(t)] = \left\langle \exp\left[\mathrm{i}\int_{-\infty}^{\infty} u(t)\Phi(t)\mathrm{d}t\right] \right\rangle \ .$$

A third manner of characterizing a stochastic process is by making use of the conditional probability space. To this end, let us split the sequence $t_1, \ldots, t_n$ into two sets $t_1, t_2, \ldots, t_r$ and $t_{r+1}, \ldots, t_n$. Given the joint density function

$$D_{\Phi}^{(r)}(\phi_1, t_1; \ldots; \phi_r, t_r)$$

of the stochastic variable Φ at points $t_1, \ldots, t_r$, the conditional density function is given by

$$\begin{aligned} &W_{\Phi}^{(n-r|r)}(\phi_n, t_n; \ldots; \phi_{r+1}, t_{r+1} | \phi_1, t_1; \ldots; \phi_r, t_r) \\ &\qquad = \frac{D_{\Phi}^{(n)}(\phi_1, t_1; \ldots; \phi_n, t_n)}{D_{\Phi}^{(r)}(\phi_1, t_1; \ldots; \phi_r, t_r)} \ . \end{aligned}$$

Statistical Properties

The mean of the product of stochastic variables $\Phi(t_1), \Phi(t_2), \ldots, \Phi(t_n)$ is calculated using the moment functions

$$\begin{aligned} M^{(n)}(t_1, \ldots, t_n) &= \langle \Phi(t_1)\Phi(t_2)\ldots\Phi(t_n) \rangle \\ &= \int_{-\infty}^{\infty} \cdots \int_{-\infty}^{\infty} \phi_1\phi_2\ldots\phi_n D_{\Phi}^{(n)}(\phi_1, \phi_2, \ldots, \phi_n)\mathrm{d}\phi_1\mathrm{d}\phi_2\ldots\mathrm{d}\phi_n \ , \end{aligned}$$

where $\phi_i = \phi(t_i)$. The function $M^{(1)}(t_1)$ is the mean of $\Phi(\omega, t)$ at time t_1. The functions $M^{(n)}(t_1, \ldots, t_n)$ are called the autocorrelation functions of $\Phi(\omega, t)$ at different times.

An important quantity is the autocorrelation time

$$\mathcal{I}^{(n)}(t_1, \ldots, t_n) = \int_0^{t_1} \cdots \int_0^{t_n} M^{(n)}(s_1, \ldots, s_n) \mathrm{d}s_1 \ldots \mathrm{d}s_n ,$$

which measures the intensity of the correlations of the stochastic process $\Phi(\omega, t)$ at different instants of time $t_1, \ldots, t_n$.

The moment functions $M^{(n)}(t_1, t_2, \ldots, t_n)$ are related to the characteristic functions $f^{(n)}$ through the formula

$$f^{(n)}(u_1, t_1; \ldots; u_n, t_n) = 1 + \sum_{s=1}^{\infty} \frac{\mathrm{i}^s}{s!} \sum_{\alpha, \ldots, \omega=1}^{n} M^{(n)}(t_\alpha, \ldots, t_\omega) u_\alpha \ldots u_\omega .$$

2.5.2 Fixed-Point Representation of Stochastic Processes

In the fixed-time representation of stochastic processes, we fix the time t and investigate each $\Phi(\omega, t_i)$ as a stochastic variable. A complementary approach is to analyse $\Phi(\omega, t)$ by fixing the point ω. We then have the fixed-point representation of the stochastic process:

$$\Phi(\omega_i, t) = \int_{-\infty}^{\infty} \Phi(\omega, t) \delta(\omega - \omega_i) \mathrm{d}\omega .$$

Here $\Phi(\omega, t)$ can be represented by a family of individual functions [57,58]:

$$\Phi(\omega, t) = [\Phi(\omega_1, t), \Phi(\omega_2, t), \ldots, \Phi(\omega_n, t)] .$$

Dynamical Properties

The stochastic differential equation (2.1) becomes

$$\dot{\Phi} \equiv \frac{\mathrm{d}}{\mathrm{d}t} \Phi = K(\Phi, \omega_i, t) . \tag{2.31}$$

Given the initial condition at t_0,

$$\Phi(\omega_i, t_0) = \Psi ,$$

equation (2.31) determines solutions for each individual point, i.e., trajectories of the stochastic process, such as the trajectories of a Brownian particle. The properties of existence, uniqueness, differentiability and continuity to be satisfied by the solutions are treated in [57,58], for example.

The dynamics of a stochastic process specified by N stochastic variables $\Phi_k(\omega_i, t), k = 1, 2, \ldots, N$, is determined by the system of ordinary differential equations

$$\begin{aligned} \dot{\Phi}_1 \equiv \frac{\mathrm{d}\Phi_1}{\mathrm{d}t} &= K_1(\Phi_1, \ldots, \Phi_N, \omega_i, t) , \\ &\vdots \\ \dot{\Phi}_N \equiv \frac{\mathrm{d}\Phi_N}{\mathrm{d}t} &= K_N(\Phi_1, \ldots, \Phi_N, \omega_i, t) . \end{aligned} \tag{2.32}$$

Statistical Properties

The statistical properties of stochastic processes can be calculated from the time-averaging procedure

$$\overline{\Phi} = \frac{1}{t} \int_0^t \Phi(\omega_i, \tau) \mathrm{d}\tau , \tag{2.33}$$

or in more general terms

$$\overline{g(\Phi)} = \frac{1}{t} \int_0^t g(\Phi(\omega_i, \tau)) \mathrm{d}\tau , \tag{2.34}$$

which is the time average of $g(\Phi)$ along a trajectory described by a given fixed point ω_i as time goes by.

2.5.3 Classification of Stochastic Processes

In this section we intend to classify the stochastic processes according to some general characteristic features present in either $\Phi(\omega, t)$ or $D_\Phi(\phi, t)$. Physically, such a classification is more concrete and comprehensible for the consideration of nonisolated systems, i.e., systems under interaction with their environment.

Nonstationary Processes

A nonstationary stochastic process $\Phi(\omega, t)$ is characterized by the explicit dependence of the solutions of the Kolmogorov stochastic equation (2.20) on time t. Physically, the nonstationarity condition means that, in general, certain probabilistic properties of a stochastic process change with time [53]. However, under quite restrictive conditions, there are physical situations that may be considered as approximately stationary. That is, if $D_\Phi(\phi, t') = D_\Phi(\phi, t' + \tau)$, for any τ, then the stochastic process is said to be stationary, i.e., independent of time: $D_\Phi(\phi, t) = D_\Phi(\phi)$.

In the fixed-time representation, the nth probability density function

$$D_\Phi^{(n)}(\phi_1, t_1; \ldots; \phi_n, t_n)$$

of a nonstationary process depends explicitly on the time values. Consequently, statistical properties like

$$\langle \Phi(t_1)\Phi(t_2)\ldots\Phi(t_n)\rangle = \int_{-\infty}^{\infty} \cdots \int_{-\infty}^{\infty} \phi_1\phi_2\ldots\phi_n D_\Phi^{(n)}(\phi_1, t_1; \ldots; \phi_n, t_n) \mathrm{d}\phi_1 \ldots \mathrm{d}\phi_n ,$$

where $\phi_i = \phi(t_i)$, depend on the absolute origin of time.

When the density functions $D_\Phi^{(n)}(\phi_1, t_1; \ldots; \phi_n, t_n)$ are invariant under a time translation

$$t_i \longmapsto t_i + \tau \quad (i = 1, 2, \ldots, n) ,$$

for arbitrary τ, the stochastic process $\Phi(\omega, t)$ is said to be (strictly) stationary [53,55,61], that is,

$$D_\Phi^{(n)}(\phi_1, t_1; \ldots; \phi_n, t_n) = D_\Phi^{(n)}(\phi_1, t_1 + \tau; \ldots; \phi_n, t_n + \tau) .$$

Setting $\tau = -t_1$, we can check that the densities $D_\Phi^{(n)}$

$$D_\Phi^{(1)}(\phi_1, t_1) = D_\Phi^{(1)}(\phi_1) ,$$
$$D_\Phi^{(2)}(\phi_1, t_1; \phi_2, t_2) = D_\Phi^{(2)}(\phi_1; \phi_2, t_2 - t_1) ,$$
$$\vdots$$
$$D_\Phi^{(n)}(\phi_1, t_1; \ldots; \phi_n, t_n) = D_\Phi^{(n)}(\phi_1, ; \ldots; \phi_n, t_n - t_1) ,$$

depend only on time differences $t_i - t_j$, where $i, j = 1, 2, \ldots, n$. Consequently, the statistical properties become

$$\langle \Phi(t_1)\rangle = \langle \Phi(t_1 + \tau)\rangle = \langle \Phi(0)\rangle = M^{(1)}(0) = \text{const.} ,$$

$$\langle \Phi(t_1)\Phi(t_2)\rangle = \langle \Phi(t_1+\tau)\Phi(t_2+\tau)\rangle = \langle \Phi(0)\Phi(t_2-t_1)\rangle = M^{(2)}(t_2-t_1) ,$$

and for the nth relation,

$$\langle \Phi(t_1) \dots \Phi(t_n) \rangle = \langle \Phi(t_1 + \tau) \dots \Phi(t_n + \tau) \rangle$$
$$= \langle \Phi(0) \dots \Phi(t_n - t_1) \rangle = M^{(n)}(t_n - t_1) \, .$$

An operational definition of stationarity is based on the assumptions

$$|\langle \Phi(t_1) \rangle| = \text{const.} < \infty \, ,$$
$$\langle \Phi^2(t_1) \rangle < \infty \, ,$$
$$\langle \Phi(t_1)\Phi(t_2) \rangle = M^{(2)}(t_2 - t_1) \, .$$

Then the stochastic process $\Phi(\omega, t)$ is said to be weakly stationary. No restriction is imposed on the density functions $D_{\Phi}^{(n)}(\phi_1, t_1; \dots ; \phi_n, t_n)$ for $n > 2$. In this context, it is useful to investigate the concept of stationary spectral density

$$S_{\Phi}(\nu) = 2 \int_{-\infty}^{\infty} \mathrm{e}^{\mathrm{i}\nu t'} \langle \Phi(0)\Phi(t') \rangle \mathrm{d}t' \, ,$$

where $t' = t_2 - t_1$ and ν is the frequency associated with the process.

Non-Gaussian Processes

A stochastic process is called Gaussian if the solutions of the Kolmogorov stochastic equation (2.20) are Gaussian functions, i.e.,

$$D_{\Phi}(\phi, t) = \left(\frac{A}{2\pi} \right)^{1/2} \exp \left(-\frac{1}{2} A\phi^2 - B\phi - \frac{B^2}{2A} \right) \, . \tag{2.35}$$

The terms $A = A(t)$ and $B = B(t)$ also depend on t. The characteristic function associated with this function is

$$f_{\Phi}(u, t) = \exp \left(-\mathrm{i}\frac{B}{A}\phi - \frac{u^2}{2A} \right) \, . \tag{2.36}$$

By calculating

$$\langle \Phi(t) \rangle = -\frac{B}{A} = \alpha_1(t) \, , \tag{2.37}$$

$$\langle (\Phi(t) - \alpha_1)^2 \rangle = \frac{1}{A} = \sigma^2(t) \, , \tag{2.38}$$

$$\langle (\Phi(t) - \alpha_1)^n \rangle = 0 \, , \qquad n \geq 3 \, , \tag{2.39}$$

we can check that the Gaussian function is completely determined by the mean α_1 and variance σ^2, that is,

$$D_\Phi(\phi,t) = \frac{1}{\sqrt{2\pi\sigma^2}} \exp\left[-\frac{(\phi-\alpha_1)^2}{2\sigma^2}\right] . \tag{2.40}$$

It should be emphasized that in general the dynamics (2.20) determines non-Gaussian solutions.

In the fixed-time representation, a Gaussian stochastic process is characterized by the fact that all the functions $D_\Phi^{(n)}(\phi_1,t_1;\ldots;\phi_n,t_n)$ are Gaussian. Of special interest are the Gaussian processes with the properties

$$\langle \Phi(t_1)\ldots\Phi(t_{2n+1})\rangle = 0 , \tag{2.41}$$

$$\langle \Phi(t_1)\ldots\Phi(t_{2n})\rangle = \sum_{r_1,\ldots,r_{2n}} \langle \Phi_{r_1}\Phi_{r_2}\rangle\langle \Phi_{r_2}\Phi_{r_3}\rangle\ldots\langle \Phi_{r_{n-1}}\Phi_{r_{2n}}\rangle . \tag{2.42}$$

The sum should be taken over all possible combinations in which one can divide the $2n$ points $t_1,\ldots,t_{2n}$ into n pairs. The number of terms is $1.3.5\ldots(n-3)(n-1)$. Equations (2.41) and (2.42) imply that we only need the density function $D_\Phi^{(2)}(\phi_1,t_1;\phi_2,t_2)$ to specify the statistical features of this type of Gaussian process. For example, let us consider the case $n=2$. We then have

$$\langle \Phi(t_1)\rangle = \langle \Phi(t_1)\Phi(t_2)\Phi(t_3)\rangle = \langle \Phi(t_1)\ldots\Phi(t_5)\rangle = 0 , \tag{2.43}$$

$$\begin{aligned} &\langle \Phi(t_1)\Phi(t_2)\Phi(t_3)\Phi(t_4)\rangle \\ &\quad = \langle \Phi_1\Phi_2\rangle\langle \Phi_3\Phi_4\rangle + \langle \Phi_1\Phi_3\rangle\langle \Phi_2\Phi_4\rangle + \langle \Phi_1\Phi_4\rangle\langle \Phi_2\Phi_4\rangle . \end{aligned} \tag{2.44}$$

Stochastic processes that do not satisfy the Gaussian conditions are called non-Gaussian. Such processes are analysed using the quasi-moment functions [53].

Neither from the mathematical point of view nor from the physical standpoint is there any deep reason to restrict the study of stochastic processes to the Gaussian case. On the contrary, physical properties are revealed with greater richness insofar as the interaction between a system and its environment is described by non-Gaussian functions [53]. From the mathematical point of view, non-Gaussian processes have been widely investigated [62,63] in a context where Gaussian processes are merely a special case.

Non-Markovian Processes

As we have just seen, in the fixed-point representation a generic stochastic process $\Phi(\omega, t)$ is determined by an infinite hierarchy of probability density functions

$$\begin{gathered} D_{\Phi}^{(1)}(\phi_1, t_1) \,, \\ D_{\Phi}^{(2)}(\phi_1, t_1; \phi_2, t_2) \,, \\ \vdots \\ D_{\Phi}^{(n)}(\phi_1, t_1; \dots; \phi_n, t_n) \,, \end{gathered}$$

or in terms of conditional density functions,

$$\begin{aligned} W^{(n-k|k)}&(\phi_n, t_n; \dots; \phi_{k+1}, t_{k+1} | \phi_1, t_1; \dots; \phi_k, t_k) \\ &= \frac{D_{\Phi}^{(n)}(\phi_1, t_1; \dots; \phi_k, t_k; \phi_{k+1}, t_{k+1}; \dots; \phi_n, t_n)}{D_{\Phi}^{(k)}(\phi_1, t_1; \dots; \phi_k, t_k)} \,. \end{aligned}$$

Let us consider the case $k = 1$. Using the compatibility condition (2.29), we obtain

$$\begin{aligned} &D_{\Phi}^{(1)}(\phi_n, t_n) \\ &= \int_{-\infty}^{\infty} \dots \int_{-\infty}^{\infty} W^{(n-1|1)}(\phi_n, t_n; \dots; \phi_2, t_2 | \phi_1, t_1) D_{\Phi}^{(1)}(\phi_1, t_1) \mathrm{d}\phi_1 \dots \mathrm{d}\phi_{n-1} \,. \end{aligned}$$

A stochastic process is said to be Markovian if one requires that [55,64]

$$\begin{aligned} W^{(n-1|1)}&(\phi_n, t_n; \dots; \phi_2, t_2 | \phi_1, t_1) \\ &= W^{(1|1)}(\phi_n, t_n | \phi_{n-1}, t_{n-1}) \dots W^{(1|1)}(\phi_2, t_2 | \phi_1, t_1) \,, \end{aligned} \tag{2.45}$$

from which one derives the relation

$$D_{\Phi}^{(1)}(\phi_n, t_n) = \int_{-\infty}^{\infty} W^{(1|1)}(\phi_n, t_n | \phi_1, t_1) D_{\Phi}^{(1)}(\phi_1, t_1) \mathrm{d}\phi_1 \,. \tag{2.46}$$

Thus a Markov process is completely characterized by $D_{\Phi}^{(1)}$ and $W^{(1|1)}$, and the hierarchy of probability density functions reduces to

$$D_{\Phi}^{(1)}(\phi_1, t_1) \tag{2.47}$$

and

$$D_{\Phi}^{(2)}(\phi_1, t_1; \phi_n, t_n) = W^{(1|1)}(\phi_n, t_n | \phi_1, t_1) D_{\Phi}^{(1)}(\phi_1, t_1) \,. \tag{2.48}$$

We have used the following general consistency relation

$$
\begin{aligned}
&W^{(1|1)}(\phi_n, t_n|\phi_1, t_1) \qquad (2.49)\\
&= \int_{-\infty}^{\infty} \dots \int_{-\infty}^{\infty} W^{(1|1)}(\phi_n, t_n|\phi_{n-1}, t_{n-1}) \dots W^{(1|1)}(\phi_2, t_2|\phi_1, t_1) \mathrm{d}\phi_{n-1} \dots \mathrm{d}\phi_2 \, ,
\end{aligned}
$$

which is the so-called Chapman–Kolmogorov equation for $n = 3$. It should be noted that the Chapman–Kolmogorov equation, derived from (2.49), is only a necessary condition to be obeyed by Markov processes. Non-Markovian processes satisfy the Chapman–Kolmogorov relation as well [55]. Equation (2.49) says that a Markov process is not conditioned by the intermediate steps between t_1 and t_n.

Stochastic processes for which the condition (2.45) is not satisfied are called non-Markovian processes. Physically, a non-Markovian description is more complete because it takes into account more details about the interaction caused by the environment at each instant of time. In contrast with this, a Markovian description depends only on the starting and finishing points. Therefore it is straightforward to conclude that 'non-Markov is the rule, Markov is the exception' [65].

It should still be remarked that the Markov condition (2.45) is defined in the fixed-time representation of stochastic processes, whereas the Kolmogorov dynamics (2.20) does not presuppose such a hypothesis. In principle, therefore, all stochastic processes are non-Markovian.

An operational definition of a Markovian stochastic process is obtained by imposing a condition on the stationary correlation function $\langle \Phi(t_1)\Phi(t_2)\rangle$, viz.,

$$
\langle \Phi(t_1)\Phi(t_2)\rangle = \frac{D}{t_\mathrm{c}} \mathrm{e}^{-(t_2-t_1)/t_\mathrm{c}} \, , \qquad t_2 > t_1 \, .
$$

where D is a constant and t_c is a characteristic time called the correlation time of the stochastic process $\Phi(t)$ at two different instants t_1 and t_2. The intensity of the correlation is calculated by means of the formula

$$
\mathcal{I}(t) = \int_0^t \langle \Phi(t_1)\Phi(t_2)\rangle \mathrm{d}t_2 = (1 - \mathrm{e}^{-t/t_\mathrm{c}})D \, .
$$

In the case $t_\mathrm{c} \to 0$, we have a punctual correlation

$$
\langle \Phi(t_1)\Phi(t_2)\rangle = D\delta(t_1 - t_2)
$$

characterizing the Markov property.

Nonlinear Processes

The stochastic differential equation (2.1) is in general nonlinear. That is, if Φ' and Φ'' are separately solutions of (2.1), then we have

$$\frac{\mathrm{d}}{\mathrm{d}t}(\Phi' + \Phi'') \neq K(\Phi', \omega, t) + K(\Phi'', \omega, t) .$$

On the other hand, the Kolmogorov stochastic equation (2.20) may be linear or nonlinear. In this book we restrict ourselves to the linear case. Therefore, any allusion to nonlinearity should be taken as being a property associated only with the differential equation (2.1). (Methods of solution of nonlinear stochastic differential equations are investigated in [56].)

Nonconservative Processes

Having defined the stochastic system (2.5) and the geometrical scenario Γ (phase space), we are able to establish a criterion characterizing the conservative or nonconservative nature of a given stochastic process. We shall see that the time behavior of a certain region $\mathcal{D}$ of phase space Γ, i.e., $\mathcal{D} \subset \Gamma$, yields the basis for such a such criterion [29]. Let

$$\mathcal{V}(\tau) = \int_{\mathcal{D}(\tau)} \mathrm{d}\Phi'_1 \ldots \mathrm{d}\Phi'_N$$

and

$$\mathcal{V}(t) = \int_{\mathcal{D}(t)} \mathrm{d}\Phi_1 \ldots \mathrm{d}\Phi_N$$

be the volumes occupied by $\mathcal{D}$ at infinitesimally distinct times $\tau = t + \Delta t$ and t, respectively. Noting that

$$\mathrm{d}\Phi_1 \ldots \mathrm{d}\Phi_N = \mathcal{J} \mathrm{d}\Phi'_1 \ldots \mathrm{d}\Phi'_N , \tag{2.50}$$

where $\mathcal{J}$ is the determinant or Jacobian

$$\mathcal{J} = \left\| \frac{\partial(\Phi'_1 \ldots \Phi'_N)}{\partial(\Phi_1 \ldots \Phi_N)} \right\|$$

of the transformation

$$(\Phi_1 \ldots \Phi_N) \longmapsto (\Phi'_1 \ldots \Phi'_N),$$

and expanding $\Phi'_i(t + \Delta t)$ around Δt to give

$$\Phi_i'(t+\Delta t) = \Phi_i(t) + \Delta t \frac{\mathrm{d}}{\mathrm{d}t}\Phi_i(t) + O[(\Delta t)^2] ,$$

we obtain

$$\mathcal{J} = 1 + \Delta t \sum_{i=1}^{N} \frac{\partial}{\partial \Phi_i} K_i + O[(\Delta t)^2] . \tag{2.51}$$

Inserting (2.51) into (2.50) and taking into account the infinitesimal character of Δt, i.e., letting $\Delta t \to 0$, it follows that

$$\frac{\mathrm{d}\mathcal{V}(t)}{\mathrm{d}t} = \int_{\mathcal{D}(t)} \left(\sum_{i=1}^{N} \frac{\partial K_i}{\partial \Phi_i} \right) \mathrm{d}\Phi_1 \ldots \mathrm{d}\Phi_N .$$

Then a dynamical stochastic system is said to be conservative, in the geometrical sense of preserving the volume during its time evolution, if and only if the divergence is null, i.e.,

$$\mathrm{div}\boldsymbol{K} = \sum_{i=1}^{N} \frac{\partial K_i}{\partial \Phi_i} = 0 ,$$

and non-conservative if and only if

$$\mathrm{div}\boldsymbol{K} = \sum_{i=1}^{N} \frac{\partial K_i}{\partial \Phi_i} \neq 0 .$$

In the case $\mathrm{div}\boldsymbol{K} < 0$, the stochastic system is said to be dissipative provided that the time average has the behavior [29]

$$\overline{\mathrm{div}\boldsymbol{K}} = \frac{1}{t} \int_0^t \mathrm{div}\boldsymbol{K} \mathrm{d}\tau < 0 .$$

Nonergodic Processes

A description of stochastic processes based on individual trajectories makes use of the time-averaging procedure

$$\overline{g(\Phi)} = \frac{1}{t} \int_0^t g(\Phi(\omega, \tau)) \mathrm{d}\tau ,$$

whereas in terms of an ensemble of trajectories, we have the so-called ensemble average

$$\langle g(\Phi) \rangle = \int_{-\infty}^{\infty} \cdots \int_{-\infty}^{\infty} g(\phi_1, \ldots, \phi_n) D_{\Phi}^{(n)}(\phi_1, t_1; \ldots; \phi_n, t_n) \mathrm{d}\phi_1 \ldots \mathrm{d}\phi_n .$$

If in the limiting case $t \to \infty$ and under certain additional conditions (e.g., $\mathrm{div}\boldsymbol{K} = 0$), one obtains the identity

$$\lim_{t\to\infty} \overline{g(\Phi)} = \lim_{t\to\infty} \left[\frac{1}{t} \int_0^t g(\Phi(\omega,\tau))\mathrm{d}\tau \right] = \langle g(\Phi) \rangle ,$$

then the stochastic process is said to be ergodic [56,58,61]. However, if such a condition is not satisfied the process is then nonergodic. In other words, there are stochastic properties inherent in the individual trajectories that cannot be revealed by an ensemble of them. On the other hand, there are ensemble properties that are not present in the individual trajectories. The breakdown of the equivalence between these two descriptions reflects Haken's synergetic principle [22]: the whole is not the mere sum of its parts; and the parts are not small copies of the whole.

Deterministic Processes

In the fixed-time representation, the stochastic processes $\Phi(\omega, t)$ characterized by Dirac delta density functions, i.e.,

$$D_\Phi^{(n)}(\phi_1, t_1; \ldots; \phi_n, t_n) = \delta(\phi_1 - \varphi_1)\delta(\phi_2 - \varphi_2)\ldots\delta(\phi_n - \varphi_n) ,$$

are said to be deterministic. Here we need only the first-order density function $D_\Phi^{(1)}(\phi_i)$. Consequently, the conditional density functions are

$$W^{(n-k|k)}(\phi_n, t_n; \ldots; \phi_{k+1}, t_{k+1}|\phi_1, t_1; \ldots; \phi_k, t_k) \\ = \delta(\phi_{k+1} - \varphi_{k+1})\ldots\delta(\phi_n - \varphi_n) .$$

The moments of the stochastic function $\Phi(\omega, t)$ give rise to the deterministic functions

$$\langle \Phi(t_1) \rangle = \varphi(t_1) ,$$

$$\vdots$$

$$\langle \Phi(t_1)\ldots\Phi(t_n) \rangle = \varphi(t_1)\ldots\varphi(t_n) .$$

There is no correlation and the variance is zero.

3 Classical Physics

3.1 Motivation: Dissipation and Fluctuation

In the previous chapter we listed some notions of the mathematical theory of stochastic processes. The present chapter aims to endow the theory with some physical content. To this end, we use a Brownian particle immersed in a fluid as our paradigmatic model of an open system exhibiting dissipation and fluctuation. We do not intend to explore all the details of the theory of Brownian motion. We choose only certain topics in order to reveal classical physics within a non-Galilean view.

We suppose the environment acts on the particle in the following ways:

- Through a friction force responsible for the dissipation of the energy of the particle. This dissipated energy spreads out within the fluid in an irreversible way.
- Through a synergetic or cooperative force exerted by the fluid particles as a whole. This force is stochastic and causes fluctuations in the physical quantities describing the Brownian particle, such as position and momentum.
- Through the initial conditions exhibiting the entanglement between the fluid and the Brownian particle.
- Through the boundary conditions specifying the thermodynamic behavior of the interaction between the Brownian particle and the fluid, so that the total system reaches either a thermal equilibrium state or a nonequilibrium stationary state.

Moreover, we assume that the physics of the environment is a kind of thermo-hydrodynamics describing equilibrium and/or non-equilibrium properties, so that the friction and diffusion coefficients present in the equations of motion of the Brownian particle can be determined by thermodynamic properties such as temperature and the Boltzmann constant, and hydrodynamic properties (e.g., viscosity of a fluid).

This chapter is organized as follows. In Sect. 3.2, we begin with the stochastic approaches due to Einstein, Smoluchowski and Rayleigh in configuration space (position or momentum space). Then in Sect. 3.3, we introduce the phase space approach (Langevin, Klein, Kramers). Section 3.4 is dedicated to the study of the deterministic limit of open systems, i.e., when the fluctuation phenomena do not influence the dissipative dynamics of the particle. Neglecting the dissipative effects, the description is made by means of the Hamilton, Lagrange and Jacobi formalisms. Finally, in Sect. 3.5, we ask how it is possible to investigate open systems as isolated systems.

3.2 Configuration Space

3.2.1 Einstein

The first theoretical model of a stochastic system was elaborated by Einstein in 1905 [36]. He considered a particle with mass m immersed in a fluid (a liquid) consisting of infinitely many particles. In general, due to the influence of the environment, there are spatial and temporal correlations which reveal the type of dynamics performed by the Brownian particle as well as the kind of geometry underlying the environment. Let the position be a stochastic process whose probabilistic nature is specified by both the function $X = X(\omega, t)$ and the probability density function $D_X(x,t)$. (For the sake of simplicity, we consider the one-dimensional case with the time t limited to the interval $0 \leq t < \infty$.)

From a mathematical point of view, the complete time evolution of this continuous stochastic process can be determined by the Kolmogorov stochastic equation (2.20):

$$\frac{\partial}{\partial t} D_X(x,t) = \sum_{s=1}^{\infty} \frac{(-1)^s}{s!} \frac{\partial^s}{\partial x^s} \left[B^{(s)}(x,t) D_X(x,t) \right] , \tag{3.1}$$

with

$$B^{(s)}(x,t) = \lim_{\Delta t \to 0} \frac{\langle [X(t+\Delta t) - X(t)]^s \rangle}{\Delta t} . \tag{3.2}$$

Nevertheless, the infinite number of terms present in (3.1) makes it intractable from the physical standpoint. In order to overcome this difficulty, we introduce the following assumption: there exists a physical time interval $\epsilon = \Delta t$ and a space interval $\xi(t, \epsilon) = \Delta X = X(t+\epsilon) - X(t)$, such that

$$\xi^3 \ll R\,, \qquad \epsilon^3 \ll Z\,.$$

With respect to R and Z, the quantities ξ and ϵ are thus considered to be infinitesimal, i.e,

$$\langle \xi^3 \rangle \to 0 \quad \Longrightarrow \quad [x(t+\epsilon) - x(t)]^3 \to 0\,. \tag{3.3}$$

Consequently, from (3.2), we obtain

$$B^{(3)}(x,t) = \lim_{\epsilon \to 0} \frac{\langle \xi^3 \rangle}{\epsilon} = 0\,, \tag{3.4}$$

$$B^{(4)}(x,t) = \lim_{\epsilon \to 0} \frac{\langle \xi^4 \rangle}{\epsilon} = 0\,. \tag{3.5}$$

According to the Pawula theorem [59], (3.5) implies that all the coefficients $B^{(n)}(x,t)$, $n = 5, 6, \ldots$, are zero. As a consequence of these arguments, the Kolmogorov stochastic equation (3.1) becomes the so-called Fokker–Planck equation in one variable [66–69]:

$$\frac{\partial D_X(x,t)}{\partial t} = -\frac{\partial}{\partial x}\left[B^{(1)}(x,t) D_X(x,t)\right] + \frac{1}{2}\frac{\partial^2}{\partial x^2}\left[B^{(2)}(x,t) D_X(x,t)\right]\,, \tag{3.6}$$

where

$$B^{(1)}(x,t) = \lim_{\epsilon \to 0} \frac{\langle \xi \rangle}{\epsilon}\,, \tag{3.7}$$

$$B^{(2)}(x,t) = \lim_{\epsilon \to 0} \frac{\langle \xi^2 \rangle}{\epsilon}\,. \tag{3.8}$$

Why must ξ and ϵ be considered as infinitesimal? What is the physical reason supporting such an interpretation? This is in fact quite a controversial question. Einstein [36] gave an operational answer (see also [37,70,71]):

> Wir führen ein Zeitintervall [ϵ] in die Betrachtung ein, welches sehr klein sei gegen die beobachtbaren Zeitintervalle, aber doch so gross, dass die in zwei aufeinanderfolgenden Zeitintervallen [ϵ] von einem Teilchen ausgeführten Bewegungen als voneinander unabhängige Ereignisse aufzufassen sind.

> We introduce a time interval [ϵ] that is very small with respect to a given time interval of observation, but large enough to ensure that the motions of the Brownian particle at two intervals of time [ϵ] can be treated as independent of each other.

This is a condition for a good observation! Despite the historical success of this assumption, Fürth [72] pointed out that the introduction of such a time interval ϵ is the weak point of the Einstein theory of Brownian motion because there is no theoretical justification for the independence of the motion in this interval.

In Chap. 4 we shall show that the true physical reason for introducing the parameters ϵ and ξ, which are infinitesimal with respect to Z and R, respectively, is related to the quantum nature of matter. [Historically, it is worth noting that the Fokker–Planck equation (3.6) was obtained early on by Fokker [66] and Planck [67] in a quantum mechanical context.]

Equilibrium Stationary State

Supposing $B^{(1)}(x,t) = 0$ and $B^{(2)}(x,t) = 2C(t)$, equation (3.6) implies the diffusion equation for the probability density function:

$$\frac{\partial}{\partial t} D_X(x,t) = C(t) \frac{\partial^2}{\partial x^2} D_X(x,t) \, . \tag{3.9}$$

Having obtained the diffusion equation (3.9), all the physics of the problem (how the Brownian particle and its surroundings are correlated) now lies in determining the diffusion coefficient $C(t)$ in terms of quantities that represent the influence of the environment (e.g., the friction coefficient $\gamma(t)$ and the absolute temperature T). Einstein [36,37] investigated the thermodynamics subjacent to Brownian motion. He then supposed that the total system (environment plus Brownian particle) was thermodynamically closed, i.e., the environment acts on the particle by means of a friction force and a thermal force so that the Brownian particle undergoes a movement characterized by a state of thermal equilibrium.

In order to find an expression for the diffusion coefficient $C(t)$, let us consider a suspension of Brownian particles subjected to an external force $F(t)$. It is assumed that the motion of the concentration $\rho(x,t)$ of these suspended particles in a diffusion process in the fluid is described by the Fick equation [34,73]:

$$\frac{\partial \rho(x,t)}{\partial t} = \frac{\partial}{\partial x} \left[C(t) \frac{\partial \rho(x,t)}{\partial x} - u(t) \rho(x,t) \right] , \tag{3.10}$$

where $C(t)$ is the same diffusion coefficient as in (3.9) and $u(t)$ is the velocity of the concentration defined in terms of the relation involving the force $F(t)$ and the friction coefficient $\gamma(t)$:

$$F(t) = -\int_0^t \gamma(t')u(t')\mathrm{d}t' \ . \tag{3.11}$$

The form of this frictional force exhibits the nonlocal character of the interaction between the Brownian particle and the fluid particles. The duration of the interaction is characterized by a correlation time $t_\mathrm{c} \neq 0$, implying that the stochastic process is non-Markovian. If we suppose that after a long time ($t \gg t_\mathrm{c}$) the concentration $\rho(x,t)$ has an asymptotic behavior, so that it no longer depends on the time t and the frictional force is instantaneous or Markovian,

$$\lim_{t_\mathrm{c}\to 0} \gamma(t') = 2\gamma\delta(t'-t),$$

where γ is a friction constant, then (3.11) implies

$$u(t) = -\frac{F(t)}{\gamma} \tag{3.12}$$

and the equation

$$C(t)\frac{\mathrm{d}\rho(x)}{\mathrm{d}x} = \frac{-F(t)}{\gamma}\rho(x) \ . \tag{3.13}$$

We now assume that $\rho(x)$ is given in terms of the Maxwell–Boltzmann exponential law

$$\rho(x) \propto \mathrm{e}^{-Fx/k_\mathrm{B}T} \ , \tag{3.14}$$

where we have used the expression for the external potential $V(x,t) = \int_{-\infty}^{x} F(t)\mathrm{d}x'$. Substituting (3.14) into (3.13), we arrive at the Einstein diffusion constant

$$C = \frac{k_\mathrm{B}T}{\gamma} \ , \tag{3.15}$$

where k_B is the Boltzmann constant and T the absolute temperature of the fluid in thermodynamic equilibrium.

If the Brownian particle is spherical and the fluid is treated as a continuous medium (that is, if the mean free path of the fluid particles is small compared with the size of the Brownian particle), then the friction constant γ may be calculated by Stokes' law in hydrodynamics:

$$\gamma = 6\pi\eta a \ , \tag{3.16}$$

where a is the radius of the Brownian particle and η the viscosity of the fluid.

It is important to emphasize that the constant nature of the diffusion coefficient (3.15) is a consequence of the Markovian property of the process. A more general relation than (3.16) is provided by the time-dependent Lorentz friction coefficient [34]

$$\gamma(t) = 6\pi a \left(1 + a\sqrt{\zeta\nu/2\eta}\right) \omega_0 \mathrm{e}^{\mathrm{i}\zeta t} , \tag{3.17}$$

where ζ is the fluid density and ν the vibrational frequency. The non-Markovian character of the friction coefficient affects the diffusion dynamics of the Brownian particle. This new feature is investigated in [74–78].

From the initial condition $D_X(x, t = 0) = \delta(x - x_0)$, the diffusion equation (3.9) provides the solution

$$D_X(x,t) = \frac{1}{\sqrt{4\pi Ct}} \exp\left[-\frac{(x-x_0)^2}{4Ct}\right] , \tag{3.18}$$

which leads to the mean square displacement

$$d^2 = \langle (X - X_0)^2 \rangle = 2Ct . \tag{3.19}$$

On the basis of (3.19), Perrin [35,79,80] was able to measure the Boltzmann constant (or Avogadro's number) by investigating the motion of Brownian particles with radius of the order of 10^{-6} m (see Fig. 3.1), thus confirming the hypothesis of the atomic structure of matter. In 1926 Perrin was awarded the Nobel prize for his experimental work.

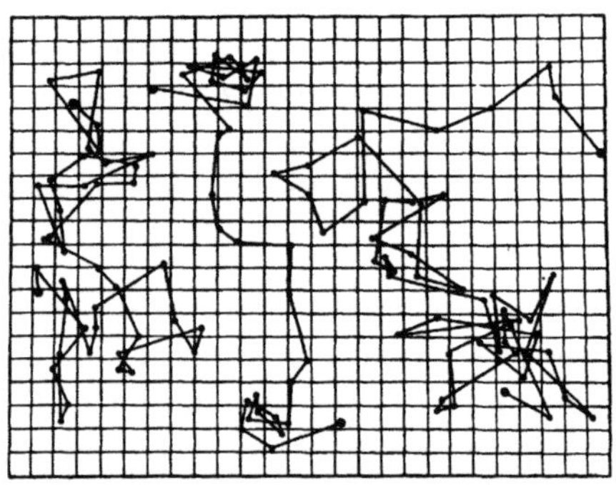

Fig. 3.1. Drawings obtained by Perrin [35] when investigating the Brownian motion of three colloidal particles of radius 0.52×10^{-6} m, as seen under the microscope. The consecutive positions were marked at intervals of 30 s, then linked by rectilinear segments having no physical reality. The mean square displacement of such segments confirms Einstein's formula (3.19)

Among several applications of the Einstein formula (3.19), it is worth mentioning the motion of cellular aggregates [81], where C is interpreted as an effective diffusion constant, measured by means of the slope of d^2 against time. Here the motion of cells with respect to the aggregate was studied from two reference frames: the center of mass of the set, and the position of the least moving cell (indicated by an asterisk in Fig. 3.2). The mean diameter of the observed cells is about 8×10^{-6} m. The two-dimensional movement of a set of five pigmented cells chosen randomly in the aggregate was recorded for 30 h at intervals of about 30 min.

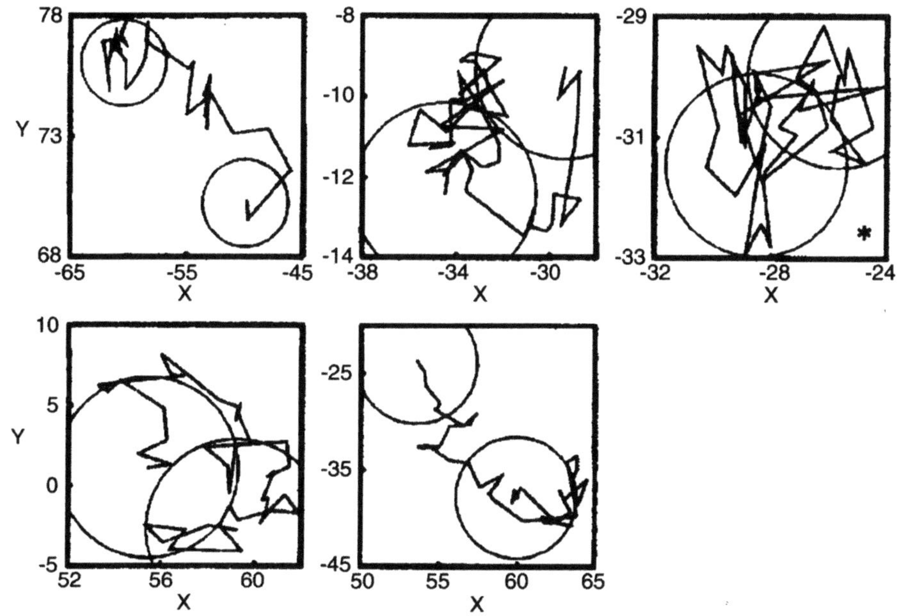

Fig. 3.2. These graphs show the initial and final positions occupied by the cells during the experiment in the center of mass reference frame. *Circles* represent the approximate size of the cells and the least moving cell is indicated by an *asterisk*. For more details see [81]

Nonequilibrium Stationary State

Now let us consider a thermodynamically open system where the environment acts on the Brownian particle by means of a friction force, a thermal force and forces of nonthermal nature, so that both environment and particle do reach a nonequilibrium stationary state characterized by the function encountered recently by Banik et al. [82]

$$\rho(x) \propto \mathrm{e}^{-Fx/k_{\mathrm{B}}\mathcal{T}} . \tag{3.20}$$

Here $\mathcal{T}$ is interpreted as the effective temperature

$$\mathcal{T} = T + \frac{\lambda}{k_B \gamma} \tag{3.21}$$

due to the presence of the nonequilibrium constant λ. When the friction constant γ is very large with respect to λ ($\gamma \gg \lambda$), we recover the thermodynamical equilibrium state (3.14). On the other hand, as $T \to 0$, or $(\lambda/k_B\gamma) \gg T$, we have a completely nonthermal environment. Following an analogous procedure to the one used to arrive at Einstein's diffusion coefficient (3.15), we set (3.20) in the time-independent Fick equation (3.13) and obtain the effective diffusion constant

$$\mathcal{C} = \frac{k_B}{\gamma}\left(T + \frac{\lambda}{k_B\gamma}\right) . \tag{3.22}$$

It is easy to check that the diffusion equation (3.9) leads to the mean square displacement given by

$$d^2 = 2\mathcal{C}t . \tag{3.23}$$

In the case of a very large friction coefficient, (3.23) coincides with (3.19), thus justifying the use of the equilibrium stationary state (3.14). From (3.22), we can determine the parameter λ as

$$\lambda = \gamma^2(\mathcal{C} - C) , \tag{3.24}$$

since the slope of d^2 vs. t can be calculated experimentally.

Anomalous Diffusion

Recently it has been observed that the mean square displacement of a diffusion process X has the following behavior [83–88]:

$$d^2 = \langle (X - X_0)^2 \rangle \propto t^\beta . \tag{3.25}$$

For $\beta > 1$, the diffusion of X is called superdiffusive, while for $\beta < 1$ it is said to be subdiffusive. In the case $\beta = 1$, the motion is normal diffusion. In experiments involving cellular aggregates, Upadhyaya et al. [89] obtained the value $\beta = 1.23 \pm 0.1$ (see Fig. 3.3).

We can derive the result (3.25) using the Fick equation (3.10). To this end, we suppose that the long-time asymptotic behavior of the friction kernel $\gamma(t)$ in (3.11) leads to the following relation between the velocity $u(t)$ and the force $F(t)$:

$$u(t) = -\frac{\beta t^{\beta-1}}{\gamma} F(t) . \tag{3.26}$$

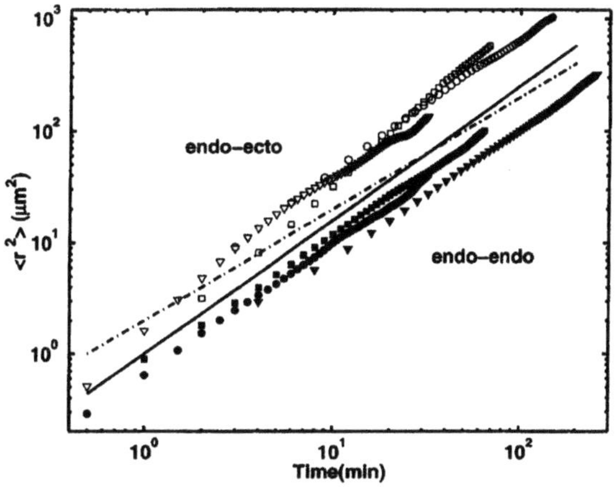

Fig. 3.3. Graph showing the two-dimensional motion of single endodermal Hydra cells in two different kinds of aggregate. The mean square displacement $\langle r^2 \rangle = \langle [x(t+t_0)-x(t_0)]^2+[y(t+t_0)-y(t_0)]^2 \rangle$ versus the time t for endodermal cells in an endodermal aggregate is represented by *filled symbols*, while the $\langle r^2 \rangle$ vs. t plot for endodermal cells in an ectodermal aggregate is represented by *open symbols*. The *solid line* has a slope of 1.2 (anomalous diffusion) while the *dashed line* has a slope of 1.0 (normal diffusion). The scale is in μm. For more details, see [89]

Using (3.13), (3.14) and (3.26), we arrive at the effective diffusion coefficient

$$\mathcal{C}(t) = \frac{k_{\mathrm{B}}T}{\gamma}\beta t^{\beta-1} , \tag{3.27}$$

found by Wang and Lung [90,91]. In (3.27), $0 < \beta < 1$ or $1 < \beta < 2$, whereas for $\beta = 1$ one obtains the Einstein diffusion coefficient (3.15). It is straightforward to check that the diffusion equation (3.9) with this time-dependent coefficient (3.27) leads to the result

$$d^2 = \langle (X - X_0)^2 \rangle = 2Ct^{\beta} . \tag{3.28}$$

The anomalous or non-Einsteinian character of the diffusive motion is exhibited by the presence of the parameter β whose values depend on the dynamics of the physical system and are associated with non-Markovian properties, i.e., with the long-range correlations present in the diffusion coefficient (3.27) via the friction coefficient in (3.11). It is still an open and fascinating issue to elaborate a theory determining the coefficients β in an a priori manner from the underlying stochastic dynamics.

Quantum Effects

At low temperatures both environment and Brownian particle could exhibit quantum effects. Besides having a thermal energy $k_B T$ activating the Brownian particle, the thermal reservoir possesses a quantum energy $\hbar\nu$, where ν is a characteristic frequency of the environment particles. Consequently, a possible quantum diffusion coefficient is given by

$$C = \frac{\hbar\nu}{\gamma} \coth\left(\frac{\hbar\nu}{k_B T}\right) , \tag{3.29}$$

reducing to the Einstein diffusion constant (3.15) in the case $k_B T \gg \nu\hbar$ or formally in the classical limit $\hbar \to 0$. Already as $k_B T \ll \nu\hbar$, i.e., as $T \to 0$, (3.29) turns out to be of an entirely quantum nature:

$$C = \frac{\hbar\nu}{\gamma} . \tag{3.30}$$

3.2.2 Smoluchowski

In 1915, assuming $B^{(1)}(x,t) = (1/m\gamma)F(x)D_X(x,t)$ and $B^{(2)}(x,t) = 2C$ in the Fokker–Planck equation (3.6), Smoluchowski [92] arrived at the generalized diffusion equation

$$\frac{\partial}{\partial t} D_X(x,t) = -\frac{1}{m\gamma}\frac{\partial}{\partial x}\Big[F(x)D_X(x,t)\Big] + C\frac{\partial^2}{\partial x^2}D_X(x,t) , \tag{3.31}$$

in terms of the external force $F(x)$ and friction constant γ. For $F(x) = 0$, the constant C is identified with the Einstein diffusion constant (3.15). For linear forces, the solutions of the Smoluchowski equation (3.31) are Gaussian functions, whereas for nonlinear forces its solutions are non-Gaussian.

The question of the validity domain of the Smoluchowski equation (3.31) was only resolved by Kramers in 1940 [93–96].

3.2.3 Rayleigh

Instead of the position, we can analyse the velocity $V = \mathrm{d}X/\mathrm{d}t$ or the momentum $P = mV$ of the Brownian particle of mass m as a stochastic process. The time evolution of the probability density function $D_P(p,t)$ associated with $P = P(\omega,t)$ is mathematically determined by the equation of motion (2.20) in the form

$$\frac{\partial D_P(p,t)}{\partial t} = \sum_{s=1}^{\infty} \frac{(-1)^s}{s!} \frac{\partial^s}{\partial p^s} \left[A^{(s)}(p,t) D_P(p,t) \right] , \tag{3.32}$$

where

$$A^{(s)}(p,t) = \lim_{\Delta t \to 0} \frac{\langle [P(t+\Delta t) - P(t)]^s \rangle}{\Delta t} . \tag{3.33}$$

As a consequence of the infinitesimal character of the parameters ξ and ϵ in (3.3), there exists a momentum $\Pi = P(t+\epsilon) - P(t)$ that has the following property of physical infinitesimality with respect to G:

$$\Pi^3 \ll G ,$$

that is,

$$\langle \Pi^3 \rangle \to 0 \quad \Longrightarrow \quad \left[p(t+\epsilon) - p(t) \right]^3 \to 0 . \tag{3.34}$$

Substituting (3.34) into (3.33), we arrive at the relations

$$A^{(3)}(p,t) = \lim_{\epsilon \to 0} \frac{\langle \Pi^3 \rangle}{\epsilon} = 0 \tag{3.35}$$

and

$$A^{(4)}(p,t) = \lim_{\epsilon \to 0} \frac{\langle \Pi^4 \rangle}{\epsilon} = 0 . \tag{3.36}$$

Applying the Pawula theorem once again [59], (3.36) and (3.32) imply

$$\frac{\partial D_P(p,t)}{\partial t} = -\frac{\partial}{\partial p} \left[A^{(1)}(p,t) D_P(p,t) \right] + \frac{1}{2} \frac{\partial^2}{\partial p^2} \left[A^{(2)}(p,t) D_P(p,t) \right] , \tag{3.37}$$

where

$$A^{(1)}(p,t) = \lim_{\epsilon \to 0} \frac{\langle \Pi \rangle}{\epsilon} , \tag{3.38}$$

$$A^{(2)}(p,t) = \lim_{\epsilon \to 0} \frac{\langle \Pi^2 \rangle}{\epsilon} . \tag{3.39}$$

Choosing $A^{(1)}(p,t) = -\gamma p$ and $A^{(2)}(p,t) = 2K = \text{const.}$, we obtain the diffusion equation in momentum space:

$$\frac{\partial D_P(p,t)}{\partial t} = \gamma \frac{\partial}{\partial p} \left[p D_P(p,t) \right] + K \frac{\partial^2}{\partial p^2} D_P(p,t) , \tag{3.40}$$

which is called the Rayleigh equation for Brownian motion [55]. In order to determine the diffusion constant K, we make use of equilibrium thermodynamics as follows.

Equilibrium Stationary State

In a stationary state, the Rayleigh equation (3.40) reads

$$\gamma p D_P(p) = -K \frac{\mathrm{d}}{\mathrm{d}p} D_P(p) \,. \tag{3.41}$$

Assuming that the probability density function $D_P(p)$ is given by the Maxwell–Boltzmann function

$$D_P(p) \propto \mathrm{e}^{-p^2/2mk_\mathrm{B}T} \,, \tag{3.42}$$

we find the expression

$$K = \gamma m k_\mathrm{B} T \tag{3.43}$$

for the diffusion constant of the particle momentum.

From the initial condition $D_P(p, t = 0) = \delta(p - p_0)$, the Rayleigh equation (3.40) with (3.43) yields the solution

$$D_P(p,t) = \left[\frac{h(t)}{\pi}\right]^{1/2} \exp\left[-h(t)(p - p_0\mathrm{e}^{-\gamma t})^2\right] \,, \tag{3.44}$$

where $h(t)$ is the function

$$h(t) = \frac{1}{2mk_\mathrm{B}T(1 - \mathrm{e}^{-2\gamma t})} \,. \tag{3.45}$$

Using (3.44), we calculate the expressions

$$\langle P(t)\rangle = p_0\mathrm{e}^{-\gamma t} \tag{3.46}$$

and

$$\langle P^2(t)\rangle = p_0^2\mathrm{e}^{-2\gamma t} + mk_\mathrm{B}T(1 - \mathrm{e}^{-2\gamma t}) \,. \tag{3.47}$$

Denoting the correlation time by $t_\mathrm{c} = 1/\gamma$, the Markovian limit $t_\mathrm{c} \to 0$ of (3.47) leads to

$$\langle P^2\rangle = mk_\mathrm{B}T \,. \tag{3.48}$$

This is the law of equipartition of energy. Combining (3.48) and (3.43), we obtain

$$\langle P^2\rangle = \frac{K}{\gamma} \,,$$

relating the fluctuation of the momentum and the friction and diffusion constants.

Nonequilibrium Stationary State

The nonequilibrium stationary state in momentum space found by Banik et al. [82] and given in terms of the effective temperature (3.21) is

$$D_P(p) \propto \mathrm{e}^{-p^2/2mk_\mathrm{B}\mathcal{T}} . \tag{3.49}$$

Substituting (3.49) into (3.41), we get the effective diffusion constant

$$K' = \gamma m k_\mathrm{B}\left(T + \frac{\lambda}{k_\mathrm{B}\gamma}\right) , \tag{3.50}$$

from which we can determine the value of λ to be

$$\lambda = \frac{1}{m}(K' - K) . \tag{3.51}$$

The Rayleigh equation is simply

$$\frac{\partial D_P(p,t)}{\partial t} = \gamma\frac{\partial}{\partial p}\Big[pD_P(p,t)\Big] + K'\frac{\partial^2}{\partial p^2}D_P(p,t) . \tag{3.52}$$

In the next section we study Brownian motion in phase space in order to exhibit the stochastic differential equations underlying the Rayleigh and Smoluchowski equations.

3.3 Phase Space

3.3.1 Langevin

In 1908 Langevin [38] suggested that the interaction of a particle with its environment may be investigated using a stochastic ordinary differential equation. Let $X = X(\omega,t)$ be the particle position and $P = P(\omega,t) = m\mathrm{d}X/\mathrm{d}t$ its physical momentum. The stochastic function $\Phi = \Phi(\omega,t)$ represents all probabilistic properties of the environment. Altogether the stochastic system comprises X, P, Φ and the corresponding joint probability density function $D_{XP\Phi}(x,p,\phi,t)$.

The motion of the Brownian particle is caused by the presence of the environment. In order to translate this causal relationship mathematically, we construct a theoretical model based on the following stochastic differential equations (the Langevin equations or generalized Langevin equations):

$$\frac{\mathrm{d}P}{\mathrm{d}t} = -\frac{\partial V(X,t)}{\partial X} - \int_0^t \gamma(t')\frac{P(t')}{m}\mathrm{d}t' + b_1\Phi(t) , \tag{3.53}$$

$$\frac{\mathrm{d}X}{\mathrm{d}t} = \frac{P}{m} + b_2 \Phi(t) , \tag{3.54}$$

where $-\partial V/\partial X$ is a force derived from an external potential $V = V(X, t)$, hence a conservative force. The term

$$\mathcal{K}(P, t) = -\int_0^t \gamma(t') \frac{P(t')}{m} \mathrm{d}t' \tag{3.55}$$

denotes a friction force depending on the time t and momentum P. It acts on the particle in a nonlocal way and is responsible for the dissipation mechanism of the particle energy ($\gamma(t) > 0$). The coefficients b_1 and b_2 depend on time t in the following way:

$$b_1(t) = \int_0^t \int_0^t \langle \Phi(t_1)\Phi(t_2) \rangle \mathrm{d}t_1 \mathrm{d}t_2 + \mathrm{e}^{-t/t_\mathrm{c}} A(t) , \tag{3.56}$$

$$b_2(t) = \int_0^t \langle \Phi(t_1) \rangle \mathrm{d}t_1 + \mathrm{e}^{-t/t_\mathrm{c}} B(t) , \tag{3.57}$$

where

$$A(t) = \sum_{k=2}^{n} \int_0^t \cdots \int_0^t \langle \Phi(t_1) \ldots \Phi(t_{2k}) \rangle \mathrm{d}t_1 \ldots \mathrm{d}t_{2k} , \tag{3.58}$$

$$B(t) = \sum_{k=1}^{n} \int_0^t \cdots \int_0^t \langle \Phi(t_1) \ldots \Phi(t_{2k+1}) \rangle \mathrm{d}t_1 \ldots \mathrm{d}t_{2k+1} . \tag{3.59}$$

The mean $\langle \cdots \rangle$ is calculated in terms of the marginal probability density function

$$D_\Phi(\phi, t) = \int_{-\infty}^{\infty} \int_{-\infty}^{\infty} D_{XP\Phi}(x, p, \phi, t) \mathrm{d}x \mathrm{d}p .$$

Both $b_1(t)$ and $b_2(t)$ measure the enviromental influence giving rise to fluctuations in the quantities describing the Brownian particle. The sum of the fluctuation parameters, $b(t) = b_1(t) + b_2(t)$, possesses all the statistical properties characterizing the stochastic force $\Phi(\omega, t)$ (seen in the fixed-time representation). We have also assumed this force to be conservative. The parameter t_c is the correlation time between our Brownian particle and the environment particles. t_c may be interpreted as the duration of a collision, for instance. In the case $t_\mathrm{c} \ll t$, or formally $t_\mathrm{c} \to 0$, so that the coefficients b_1 and b_2 turn out to be constant, the stochastic system (3.53) and (3.54) is then said to be Markovian, whereas for $t_\mathrm{c} \neq 0$, the system is non-Markovian.

It should be pointed out that the coefficients b_1 and b_2 possess the same phenomenological status as the mass m of the particle. Moreover, we do not make any deeper consideration concerning the physical origin of the forces present in (3.53) and (3.54). In order to reveal the physical relevance of our coefficients b_1 and b_2, let us examine two examples.

Example 1: Free Particle $V = 0$ without Friction Force

Let us consider a free Brownian particle $V = 0$. During a time τ, we imagine that the fluctuating force is very large with respect to the friction force. An approximate description of this situation is thus given by the equations ($0 \leq t \leq \tau$)

$$\frac{\mathrm{d}P}{\mathrm{d}t} = b_1(t)\Phi(t) , \tag{3.60}$$

$$\frac{\mathrm{d}X}{\mathrm{d}t} = \frac{P}{m} + b_2(t)\Phi(t) . \tag{3.61}$$

We also assume that the stochastic force $\Phi(t)$ has the statistical properties

$$\langle \Phi(t_1) \rangle = \frac{D}{t_\mathrm{c}} \mathrm{e}^{-t_1/t_\mathrm{c}} , \tag{3.62}$$

$$\langle \Phi(t_1)\Phi(t_2) \rangle = \frac{D}{t_\mathrm{c}} \mathrm{e}^{-(t_1+t_2)/2t_\mathrm{c}} \delta(t_1 - t_2) , \tag{3.63}$$

$$\langle \Phi(t_1) \ldots \Phi(t_{2k+1}) \rangle = 0 \quad (k = 1, 2, \ldots) , \tag{3.64}$$

$$\langle \Phi(t_1) \ldots \Phi(t_{2k}) \rangle = 0 \quad (k = 2, 3, \ldots) . \tag{3.65}$$

Consequently, our coefficients b_1 and b_2 in (3.60) and (3.61) are

$$b_1(t) = b_2(t) = D\left(1 - \mathrm{e}^{-t/t_\mathrm{c}}\right) . \tag{3.66}$$

In the limit $t_\mathrm{c} \to 0$, that is, $t_\mathrm{c} \ll t$, the motion of our Brownian particle becomes Markovian.

With the initial condition $P_0 = P(0) = 0$, (3.60) reads

$$P(\tau) = \int_0^\tau b_1(t')\Phi(t')\mathrm{d}t' . \tag{3.67}$$

Averaging (3.67) and using (3.62), we obtain

$$\langle P \rangle = D^2 \left(\frac{1}{2} \mathrm{e}^{-2\tau/t_\mathrm{c}} - \mathrm{e}^{-\tau/t_\mathrm{c}} + \frac{1}{2} \right) , \tag{3.68}$$

while an analogous procedure for P^2 leads to

$$\langle P^2 \rangle = D^3 \left(\mathrm{e}^{-2\tau/t_\mathrm{c}} - \mathrm{e}^{-\tau/t_\mathrm{c}} - \frac{1}{3}\mathrm{e}^{-3\tau/t_\mathrm{c}} + \frac{1}{3} \right) . \tag{3.69}$$

On the other hand, from (3.61), we have the equation ($X_0 = 0$)

$$X(\tau) = \frac{1}{m} \int_0^\tau \mathrm{d}s \int_0^s b_1(t')\Phi(t')\mathrm{d}t' + \int_0^\tau b_2(t')\Phi(t')\mathrm{d}t' , \tag{3.70}$$

which implies

$$\langle X \rangle = \frac{D^2 t_\mathrm{c}}{m} \left(\mathrm{e}^{-\tau/t_\mathrm{c}} - \frac{1}{4}\mathrm{e}^{-2\tau/t_\mathrm{c}} - \frac{3}{4} \right) + \frac{D^2}{2} \left(1 - \mathrm{e}^{-\tau/t_\mathrm{c}} \right)^2 + \frac{D^2 \tau}{2m} \tag{3.71}$$

and

$$\langle X^2 \rangle = I_1 + I_2 + I_3 , \tag{3.72}$$

where

$$I_1 = \frac{D^3 t_\mathrm{c}^2}{m^2} \left[(\mathrm{e}^{-\tau/t_\mathrm{c}} - 1) + \frac{1}{4}(\mathrm{e}^{-2\tau/t_\mathrm{c}} - 1) - \frac{1}{27}(\mathrm{e}^{-3\tau/t_\mathrm{c}} - 1) \right] - \frac{11}{18} \frac{D^3 t_\mathrm{c}}{m^2} \tau , \tag{3.73}$$

$$I_2 = \frac{2D^3}{m} \left[t_\mathrm{c} \left(-\frac{1}{2}\mathrm{e}^{-2\tau/t_\mathrm{c}} + \mathrm{e}^{-\tau/t_\mathrm{c}} + \frac{1}{9}\mathrm{e}^{-3\tau/t_\mathrm{c}} - \frac{11}{18} \right) + \frac{\tau}{3} \right] , \tag{3.74}$$

$$I_3 = D^3 \left(-\mathrm{e}^{-\tau/t_\mathrm{c}} + \mathrm{e}^{-2\tau/t_\mathrm{c}} - \frac{1}{3}\mathrm{e}^{-3\tau/t_\mathrm{c}} + \frac{1}{3} \right) . \tag{3.75}$$

Multiplying (3.67) and (3.70), we arrive at

$$\langle XP \rangle = \frac{D^3 t_\mathrm{c}}{m} \left(\frac{25}{18} + \frac{\mathrm{e}^{-3\tau/t_\mathrm{c}}}{9} - \frac{\mathrm{e}^{-2\tau/t_\mathrm{c}}}{2} - \mathrm{e}^{-\tau/t_\mathrm{c}} \right) + \frac{D^3}{3m}\tau + D^3 \left(\frac{1 - \mathrm{e}^{-3\tau/t_\mathrm{c}}}{3} - \mathrm{e}^{-\tau/t_\mathrm{c}} + \mathrm{e}^{-2\tau/t_\mathrm{c}} \right) . \tag{3.76}$$

The constant D in (3.66) is determined by requiring that the law of energy equipartition is valid in the Markovian limit $t_\mathrm{c} \to 0$. Equation (3.69) then yields

$$\langle P^2 \rangle = \frac{D^3}{3} = m k_\mathrm{B} T \quad \Longrightarrow \quad D = (3 m k_\mathrm{B} T)^{1/3} . \tag{3.77}$$

Our diffusion coefficients in (3.66) are then

$$b_1(t) = b_2(t) = (3mk_\mathrm{B}T)^{1/3}\left(1 - \mathrm{e}^{-t/t_\mathrm{c}}\right) ,$$

expressed in terms of the Boltzmann constant k_B, absolute temperature T of the environment, and mass m of the Brownian particle.

Assuming $\tau = t/\gamma$, where γ is the friction coefficient and taking the limit $t_\mathrm{c} \to 0$, equations (3.71) and (3.72) become

$$\langle X \rangle = -\frac{3D^2}{2m}\frac{t}{\gamma} , \tag{3.78}$$

$$\langle X^2 \rangle = \frac{2D^3}{3m}\frac{t}{\gamma} . \tag{3.79}$$

Calculating the mean square displacement

$$d^2 = \langle X^2 \rangle - \langle X \rangle^2 = \frac{2D^3}{3m}\frac{t}{\gamma} - \frac{9D^4}{4m^2}\frac{t^2}{\gamma^2} , \tag{3.80}$$

which establishes the relation between the fluctuation of the position around its average and the friction constant, we arrive at the Einstein formula

$$d^2 \approx \frac{2k_\mathrm{B}T}{\gamma}t , \tag{3.81}$$

by considering the case of large viscosity coefficient $\gamma^{-2} \ll 1$ in (3.80). In the Markovian limit, we can also check that the momentum and position of the Brownian particle are correlated:

$$\langle XP \rangle = \frac{D^3}{3m}\frac{t}{\gamma} + \frac{D^3}{3} . \tag{3.82}$$

Example 2: Free Particle $V = 0$ with Friction Force

The next example concerns the Langevin equation for a free Brownian particle subjected to a frictional force [38,97]:

$$\frac{\mathrm{d}P}{\mathrm{d}t} = -\alpha P + b_1\Phi(t) , \tag{3.83}$$

$$\frac{\mathrm{d}X}{\mathrm{d}t} = \frac{P}{m} + b_2\Phi(t) , \tag{3.84}$$

where $\alpha = \gamma/m$. Here we have used $\gamma(t') = \gamma\delta(t'-t)$ in (3.55). Because the stochastic force $\Phi(t)$ has the statistical properties

$$\langle \Phi(t) \rangle = 0 \,, \tag{3.85}$$

$$\langle \Phi(t_1)\Phi(t_2) \rangle = 2C\delta(t_1 - t_2) \,, \tag{3.86}$$

$$\langle \Phi(t_1) \ldots \Phi(t_{2k+1}) \rangle = 0 \quad (k = 1, 2, \ldots) \,, \tag{3.87}$$

$$\langle \Phi(t_1) \ldots \Phi(t_{2k}) \rangle = 0 \quad (k = 1, 2, \ldots) \,, \tag{3.88}$$

the coefficient b_2 in (3.84) is zero. In addition, as the stochastic force Φ is strictly stationary, we define b_1 as

$$b_1 = \int_0^t \langle \Phi(t_1)\Phi(t_2) \rangle \mathrm{d}t_1 = C \,. \tag{3.89}$$

From (3.83–3.89), we obtain the quantities

$$\langle \Delta P \rangle = P_0 \left(\mathrm{e}^{-\alpha t} - 1 \right) \,, \tag{3.90}$$

$$\langle (\Delta P)^2 \rangle = P_0^2 \left(\mathrm{e}^{-2\alpha t} - 2\mathrm{e}^{-\alpha t} + 1 \right) + \frac{C^3}{2\alpha} \left(1 - \mathrm{e}^{-2\alpha t} \right) \,, \tag{3.91}$$

$$\langle \Delta X \rangle = \frac{P_0}{\alpha m} \left(1 - \mathrm{e}^{-\alpha t} \right) \,, \tag{3.92}$$

$$\begin{aligned} \langle (\Delta X)^2 \rangle = {} & \frac{P_0^2}{\alpha^2 m^2} \left(\mathrm{e}^{-2\alpha t} - 2\mathrm{e}^{-\alpha t} + 1 \right) \\ & + \frac{C^3}{2\alpha^3 m^2} \left(3\mathrm{e}^{-\alpha t} + \mathrm{e}^{\alpha t} - \mathrm{e}^{-2\alpha t} - 3 \right) \,, \end{aligned} \tag{3.93}$$

$$\begin{aligned} \langle \Delta X \Delta P \rangle = {} & -\frac{P_0^2}{\alpha m} \left(\mathrm{e}^{-2\alpha t} - 2\mathrm{e}^{-\alpha t} + 1 \right) \\ & - \frac{C^3}{2\alpha^2 m} \left(\mathrm{e}^{-2\alpha t} - \frac{1}{3}\mathrm{e}^{-4\alpha t} - \frac{2}{3}\mathrm{e}^{-\alpha t} \right) \,. \end{aligned} \tag{3.94}$$

We note that the correlation time t_c is the reciprocal of α: $t_\mathrm{c} = m/\gamma$. We can determine the value of C if, in the Markovian limit $t_\mathrm{c} \to 0$, the theorem of energy equipartition is satisfied, that is,

$$\langle P^2 \rangle = \frac{C^3}{2\alpha} = m k_\mathrm{B} T \quad \Longrightarrow \quad C = (2\gamma k_\mathrm{B} T)^{1/3} \,. \tag{3.95}$$

As $C = b_1$ in (3.89), we find the fluctuation coefficient

$$b_1 = (2\gamma k_\mathrm{B} T)^{1/3} \,, \tag{3.96}$$

in terms of thermodynamic quantities (k_{B} and T) and the friction constant γ.

Equations (3.93) and (3.92) lead to the mean square displacement or the fluctuation in the position [34,98–100]

$$\begin{aligned} d^2 &= \langle(\Delta X)^2\rangle - \langle(\Delta X)\rangle^2 \\ &= \frac{k_{\mathrm{B}}T}{\gamma^2}\left(-3m + 4m\mathrm{e}^{-\alpha t} + 2\gamma t - m\mathrm{e}^{-2\alpha t}\right) . \end{aligned} \tag{3.97}$$

In the limit $t_{\mathrm{c}} \to 0$, i.e., in the case of a very large friction constant, this reduces to the well-known Einstein result (3.19):

$$d^2 \approx \frac{2k_{\mathrm{B}}T}{\gamma} t . \tag{3.98}$$

From (3.94), we note that there is no correlation between X and P in the Markovian limit $t_{\mathrm{c}} \to 0$:

$$\langle XP\rangle = 0 . \tag{3.99}$$

Besides reproducing the same results as Einstein's theory, the Langevin approach explains the connection between the stochastic ordinary differential equation and the equation of motion for the probability density function. This question was solved by Klein [101] in 1922 and independently by Kramers in 1940 [93].

3.3.2 Klein and Kramers

The propagator $W(x, p, t'|x', p', t)$ present in the Einstein identity in phase space, viz.,

$$D_{XP}(x, p, t') = \int_{-\infty}^{\infty}\int_{-\infty}^{\infty} W(x, p, t'|x', p', t) D_{XP}(x', p', t)\mathrm{d}x'\mathrm{d}p' , \tag{3.100}$$

is related to the characteristic function in terms of the increments $\xi = \Delta X = X(t+\epsilon) - X(t)$ and $\Pi = \Delta P = P(t+\epsilon) - P(t)$ as follows [53]:

$$\langle \mathrm{e}^{\mathrm{i}(u\Delta X + r\Delta P)}\rangle = \int_{-\infty}^{\infty}\int_{-\infty}^{\infty} \mathrm{e}^{\mathrm{i}(u\Delta X + r\Delta P)} W(x, p, , t'|x', p', t)\mathrm{d}x'\mathrm{d}p' . \tag{3.101}$$

Using the inverse of (3.101)

$$W(x,p,t'|x',p',t) = \frac{1}{2\pi}\int_{-\infty}^{\infty}\int_{-\infty}^{\infty} \mathrm{e}^{-\mathrm{i}(u\Delta x + r\Delta p)}\langle \mathrm{e}^{\mathrm{i}(u\Delta X + r\Delta P)}\rangle \mathrm{d}u\mathrm{d}r \tag{3.102}$$

and the expansion

$$\langle \mathrm{e}^{\mathrm{i}(u\Delta X + r\Delta P)}\rangle = \sum_{s=0}^{\infty}\frac{\mathrm{i}^s}{s!}\langle (u\Delta X + r\Delta P)^s\rangle , \tag{3.103}$$

and following an analogous procedure to the one used to arrive at (3.6) and (3.37) (infinitesimal nature of the parameters ϵ, ξ and Π, Pawula's theorem), we obtain the Fokker–Planck equation in phase space [93,101]:

$$\begin{aligned}&\frac{\partial D_{XP}}{\partial t} \\ &= \left(-\frac{\partial}{\partial x}A_1 - \frac{\partial}{\partial p}A_2 + \frac{1}{2}\frac{\partial^2}{\partial x^2}B_{11} + 2\frac{\partial^2}{\partial x \partial p}B_{12} + \frac{1}{2}\frac{\partial^2}{\partial p^2}B_{22}\right)D_{XP} ,\end{aligned} \tag{3.104}$$

where the coefficients are given by

$$A_1 = A_1(x,p,t) = \lim_{\epsilon\to 0}\frac{\langle \Delta X\rangle}{\epsilon} , \tag{3.105}$$

$$A_2 = A_2(x,p,t) = \lim_{\epsilon\to 0}\frac{\langle \Delta P\rangle}{\epsilon} , \tag{3.106}$$

$$B_{11} = B_{11}(x,p,t) = \lim_{\epsilon\to 0}\frac{\langle (\Delta X)^2\rangle}{\epsilon} , \tag{3.107}$$

$$B_{22} = B_{22}(x,p,t) = \lim_{\epsilon\to 0}\frac{\langle (\Delta P)^2\rangle}{\epsilon} , \tag{3.108}$$

$$B_{12} = B_{12}(x,p,t) = \lim_{\epsilon\to 0}\frac{\langle (\Delta X)(\Delta P)\rangle}{\epsilon} . \tag{3.109}$$

In order to construct the Fokker–Planck equation in phase space (3.104), we start with the Langevin equations

$$\frac{\mathrm{d}P}{\mathrm{d}t} = -\frac{\partial V(X,t)}{\partial X} + \mathcal{K}(P,t) + b_1(t)\Phi(t) , \tag{3.110}$$

$$\frac{\mathrm{d}X}{\mathrm{d}t} = \frac{P}{m} + b_2(t)\Phi(t) , \tag{3.111}$$

and calculate the coefficients (3.105–3.109) from the equations

$$\Delta P = P(t+\epsilon) - P(\epsilon)$$
$$= \int_t^{t+\epsilon} \left[-\frac{\partial V(X,t')}{\partial X} + \mathcal{K}(P,t') + b_1(t')\Phi(t') \right] \mathrm{d}t' \ , \quad (3.112)$$

$$\Delta X = X(t+\epsilon) - X(\epsilon) = \frac{P}{m}\epsilon + \int_t^{t+\epsilon} b_2(t')\Phi(t')\mathrm{d}t' \ , \quad (3.113)$$

taking into account the statistical properties of the stochastic force Φ encapsulated in our coefficients $b_1(t)$ and $b_2(t)$. We thus have a method for constructing the Fokker–Planck equation (3.104) applicable to non-linear, non-Gaussian, and non-Markovian stochastic processes. In order to exhibit this technique we consider two cases below.

However, before doing so, we would like to emphasize that in order to arrive at the Fokker–Planck equation (3.104), we must use the infinitesimal nature of a certain time interval ϵ. Kramers [93] explains the need for this parameter in the following way:

> A theory of Brownian motion on the Einstein pattern can be set up if there exists a range of time intervals $[\epsilon]$ which has the following properties: On the one hand $[\epsilon]$ must be so short, that the change of velocity suffered in the course of $[\epsilon]$ may be considered as very small; on the other hand $[\epsilon]$ must be so large, that the chance for $[\Phi]$ to take a given value at the time $[t+\epsilon]$ is independent of the value which $[\Phi]$ possessed at the time t.

This is an operational condition ensuring that the motion of the Brownian particle is Markovian. Below we show that this argument is not entirely correct, since we can also derive non-Markovian equations within the general framework of the Fokker–Planck equations (3.104). Once again, we point out that the true reason for the infinitesimality of the parameters ϵ, ξ, Π is connected with the quantum nature of matter (see Chap. 4).

Example 1: Non-Markovian Systems in Phase Space

Initially, we assume that the stochastic force $\Phi(t)$ has the same properties as in (3.62–3.65). By calculating the coefficients (3.105–3.109) on the basis of (3.110–3.113), we arrive at

$$A_1(p,t) = \frac{p}{m} + \frac{D^2}{m}\left(\mathrm{e}^{-t/t_c} - \frac{1}{2}\mathrm{e}^{-2t/t_c} \right) , \quad (3.114)$$

$$A_2(x,p,t) = -\frac{\partial}{\partial x}V(x,t) + \mathcal{K}(p,t) \ , \quad (3.115)$$

$$B_{11}(t) = \frac{2D^3}{m}\left(\mathrm{e}^{-t/t_c} - \mathrm{e}^{-2t/t_c} + \frac{1}{3}\mathrm{e}^{-3t/t_c}\right) \tag{3.116}$$

$$+\frac{D^3 t_c}{m^2}\left(-\mathrm{e}^{-t/t_c} + \frac{1}{2}\mathrm{e}^{-2t/t_c} - \frac{1}{9}\mathrm{e}^{-3t/t_c}\right) ,$$

$$B_{22} = 0 , \tag{3.117}$$

$$B_{12}(t) = \frac{D^3}{m}\left(\mathrm{e}^{-t/t_c} - \mathrm{e}^{-2t/t_c} + \frac{1}{3}\mathrm{e}^{-3t/t_c}\right) . \tag{3.118}$$

The Fokker–Planck equation in phase space is therefore

$$\frac{\partial}{\partial t}D_{XP} = \left(-A_1\frac{\partial}{\partial x} - \frac{\partial}{\partial p}A_2 + \frac{B_{11}}{2}\frac{\partial^2}{\partial x^2} + 2B_{12}\frac{\partial^2}{\partial x \partial p}\right) D_{XP} . \tag{3.119}$$

Due to the nonlinearity of the external potential V and the friction force $\mathcal{K}$, the solutions of (3.119) are in general non-Gaussian functions.

For the case $t_c \to 0$, our non-Markovian Fokker–Planck equation (3.119) reduces to the deterministic Liouville equation

$$\frac{\partial}{\partial t}D_{XP} = -\frac{p}{m}\frac{\partial}{\partial x}D_{XP} + \frac{\partial}{\partial p}\left[\frac{\partial V(x,t)}{\partial x} - \mathcal{K}(p,t)\right] D_{XP} . \tag{3.120}$$

Adelman [102] and Mazo [34,103] have obtained an equation similar to our (3.119). It should be noted, however, that the Adelman–Mazo method for obtaining non-Markovian Fokker–Planck equations only works for linear forces [82,104–107]. Specifically, Adelman introduces the nonlinear force $-\partial V/\partial x$ by hand in [102].

Example 2: Markovian Systems in Phase Space

Let us consider the Markovian stochastic system (b_1 is a constant and $b_2 = 0$) described by

$$\frac{\mathrm{d}P}{\mathrm{d}t} = -\frac{\partial V(X,t)}{\partial X} + \mathcal{K}(P,t) + b_1\Phi(t) , \tag{3.121}$$

$$\frac{\mathrm{d}X}{\mathrm{d}t} = \frac{P}{m} . \tag{3.122}$$

Using the stochastic properties (3.85–3.88) for $\Phi(t)$, we construct the Fokker–Planck equation by calculating the coefficients

$$A_1 = \frac{p}{m} , \tag{3.123}$$

$$A_2 = -\frac{\partial}{\partial x} V(x,t) + \mathcal{K}(p,t) , \tag{3.124}$$

$$B_{11} = 0 , \tag{3.125}$$

$$B_{12} = 0 , \tag{3.126}$$

$$B_{22} = b_1^3 . \tag{3.127}$$

Inserting (3.123–3.127) into (3.104), we obtain the Markovian Fokker–Planck equation in phase space:

$$\frac{\partial}{\partial t} D_{XP} = -\frac{p}{m}\frac{\partial}{\partial x} D_{XP} + \frac{\partial}{\partial p}\left[\frac{\partial V}{\partial x} - \mathcal{K}(p,t)\right] D_{XP} + b_1^3 \frac{\partial^2}{\partial p^2} D_{XP} , \tag{3.128}$$

which generally generates non-Gaussian processes.

The book by Risken [68] is entirely devoted to solutions to and applications of this equation (3.128), obtained early on by Klein [101] and independently by Kramers [93].

The diffusion coefficient b_1^3 is determined as follows. Using the marginal density function

$$D_P(p,t) = \int_{-\infty}^{\infty} D_{XP}(p,x,t)\mathrm{d}x \tag{3.129}$$

in (3.128) for the case $V = 0$ and $\mathcal{K} = -\gamma p$, we identify b_1^3 with the diffusion constant (3.43) of the Rayleigh equation (3.40):

$$b_1^3 = \gamma m k_{\mathrm{B}} T . \tag{3.130}$$

When the frictional constant is very large, the Smoluchowski equation (3.31) can be derived in configuration space from the Fokker–Planck equation in phase space (3.128) [93–96]. This task was first rigorously performed by Wilemski only in 1976 [95].

It is easy to check that the stochastic differential equation associated with the Smoluchowski equation (3.31) is given by [100]

$$\frac{\mathrm{d}X}{\mathrm{d}t} = -\frac{1}{\gamma}\frac{\partial V}{\partial X} + \frac{(\gamma k_{\mathrm{B}} T)^{1/3}}{\gamma}\Phi(t) , \tag{3.131}$$

where the stochastic function Φ has the properties

$$\langle \Phi(t) \rangle = 0 ,$$

$$\langle \Phi(t_1)\Phi(t_2) \rangle = 2(\gamma k_{\mathrm{B}} T)^{1/3}\delta(t_1 - t_2) ,$$

while the equation

$$\frac{\mathrm{d}P}{\mathrm{d}t} = -\gamma p + (\gamma m k_{\mathrm{B}} T)^{1/3} \Phi(t) ,$$

with

$$\langle \Phi(t) \rangle = 0 ,$$

$$\langle \Phi(t_1) \Phi(t_2) \rangle = 2(\gamma m k_{\mathrm{B}} T)^{1/3} \delta(t_1 - t_2) ,$$

generates the Rayleigh equation (3.40).

Above we have discussed some characteristic features of the theory of Brownian motion. Important topics, such as the fractal geometry underlying Brownian trajectories, as well as chaotic, quantum and relativistic properties, still stand as terra incognita which remains to be explored. Indeed, we must recognize that 'Brownian motion, understood for almost one hundred years, still has some surprises for us, and some interesting questions, both experimental and theoretical' [34].

In the next section we survey the question of the deterministic limit of stochastic systems, i.e., the limit where, insofar as the direct influence of the environment can be neglected, the dynamics of our particle starts to exhibit deterministic properties.

3.4 Deterministic Systems

3.4.1 Newton

The fluctuation coefficients b_1 and b_2 introduced in the stochastic differential equations (3.53) and (3.54) allow one to reproduce well-known results in the theory of Brownian motion, such as the Einstein formula (3.19) via (3.60–3.65) or (3.83–3.88). Moreover, using b_1 and b_2, we are able to obtain the non-Markovian Fokker–Planck equation (3.119) for nonlinear forces. We now want to show that they are also responsible for the deterministic limit of stochastic processes.

There are physical situations where the stochastic force is not important when analysing the movement of the particle. In other words, the stochastic system can be treated as deterministic. Formally, we take the deterministic limit

$$b_1 \to 0 , \qquad b_2 \to 0 , \tag{3.132}$$

in the stochastic differential equations (3.53) and (3.54), which reduce to the Newton equations with dissipation [29,108]:

$$\frac{\mathrm{d}P}{\mathrm{d}t} = -\frac{\partial}{\partial X} V(X,t) - \mathcal{K}(P,t) , \tag{3.133}$$

$$\frac{dX}{dt} = \frac{P}{m} \,. \tag{3.134}$$

On the other hand, the deterministic limit of the Fokker–Planck equation (3.119) or (3.128) is the dissipative Liouville equation

$$\frac{\partial}{\partial t} D_{XP} = -\frac{P}{m}\frac{\partial}{\partial X} D_{XP} + \frac{\partial}{\partial P}\left[\frac{\partial}{\partial X} V(X,t) - \mathcal{K}(P,t)\right] D_{XP} \,. \tag{3.135}$$

[It is important to bear in mind that this Liouville equation (3.135) is valid within the infinitesimality domain defined by (3.3) and (3.34).] It is interesting to observe that, on the basis of the stochastic approach, we have obtained the Newtonian equations (3.133) and (3.134) and the Liouville equation (3.135) at the same time. Historically, however, there was an interval of about 150 years between the two equations: Newton introduced his equations in 1687 (Philosophie Naturalis Principia Mathematica), whereas Liouville produced his in 1838 [J. Liouville: J. Mathém. Pures Appl. **3**, 342 (1838)].

Although there is no direct environmental influence upon the particle, stochastic features are indirectly present through the friction term $\mathcal{K}(P,t)$, responsible for dissipation of mechanical energy (kinetic energy plus potential energy) and specification of initial and boundary conditions for solving the dynamical equations (3.133–3.135) (see Fig. 3.4).

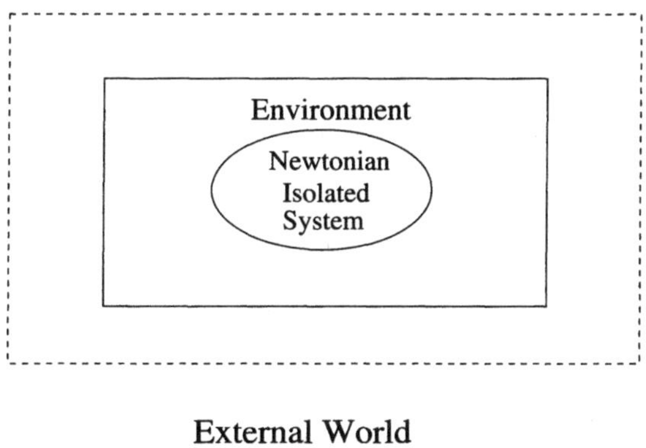

Fig. 3.4. As a consequence of the isolation process, the environment plays only a passive role in the deterministic limit: it absorbs the energy dissipated by the Newtonian isolated system

3.4.2 Hamilton

Now let us suppose that the dissipative force $\mathcal{K}(p,t)$ is absent from the Newtonian deterministic system (3.133) and (3.134) and that the initial and boundary conditions are always specified with probability one. We then obtain the conservative system [109]

$$\frac{\mathrm{d}p}{\mathrm{d}t} = -\frac{\partial V(x,t)}{\partial x} , \tag{3.136}$$

$$\frac{\mathrm{d}x}{\mathrm{d}t} = \frac{p}{m} , \tag{3.137}$$

after averaging (3.133) and (3.134) according to

$$\langle f(X,P)\rangle = \int_{-\infty}^{\infty}\int_{-\infty}^{\infty} f(x',p')\delta(x'-x)\delta(p'-p)\mathrm{d}x'\mathrm{d}p' = f(x,p) . \tag{3.138}$$

In the Hamiltonian formulation of classical mechanics, the Newtonian equations (3.136) and (3.137) are replaced by the Hamilton equations [5,109]

$$\frac{\mathrm{d}p}{\mathrm{d}t} = -\frac{\partial H}{\partial x} , \tag{3.139}$$

$$\frac{\mathrm{d}x}{\mathrm{d}t} = \frac{\partial H}{\partial p} , \tag{3.140}$$

where the Hamiltonian function $H = H(x,p,t)$ has the form

$$H(x,p,t) = \frac{p^2}{2m} + V(x,t) \tag{3.141}$$

and is interpreted physically as the total energy of the particle. The Hamiltonian system (3.139) and (3.140) is geometrically conservative (its divergence is zero), even though the mechanical energy is not conserved. In the case $H = H(x,p)$, we have the highly idealistic situation of a physical system completely isolated from its surroundings (see Fig. 3.5).

It is important to note that the Hamiltonian systems (3.139) and (3.140) constitute a very particular class of Newtonian system (3.133) and (3.134).

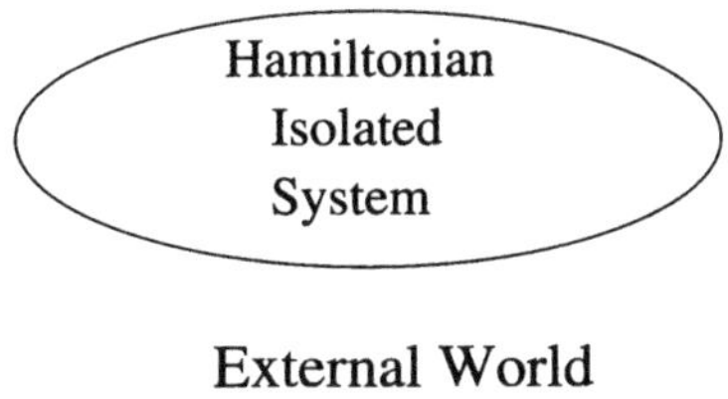

Fig. 3.5. A physical system completely isolated from its surroundings

3.4.3 Lagrange

Starting from the Hamiltonian (3.141) and performing the Legendre transform

$$L(x, \dot{x}, t) = p\dot{x} - H(p, x, t) \ , \tag{3.142}$$

we arrive at the Lagrangian formulation of classical mechanics [5,108, 109]:

$$\frac{\mathrm{d}}{\mathrm{d}t}\left(\frac{\partial L}{\partial \dot{x}}\right) = \frac{\partial L}{\partial x} \ . \tag{3.143}$$

The Lagrangian function is given by

$$L(x, \dot{x}, t) = \frac{m\dot{x}^2}{2} - V(x, t) \ . \tag{3.144}$$

The integral of the Lagrangian function $L(x, \dot{x}, t)$ along a trajectory Υ connecting the points (x', t') and (x, t) has physical significance and is called the action function [5,108,109]:

$$S(x, t) = \int_{\Upsilon} L(x, \dot{x}, t)\mathrm{d}t \ . \tag{3.145}$$

3.4.4 Jacobi

The differential form of (3.145) for a fixed initial point reads

$$\mathrm{d}S(x, t) = L(x, \dot{x}, t)\mathrm{d}t = p\mathrm{d}x - H\mathrm{d}t \ , \tag{3.146}$$

after using the Legendre transform (3.142). Noting that

$$\mathrm{d}S = \frac{\partial S}{\partial x}\mathrm{d}x + \frac{\partial S}{\partial t}\mathrm{d}t \ ,$$

equation (3.146) implies

$$p = \frac{\partial S}{\partial x} \tag{3.147}$$

and

$$H(p,x,t) = -\frac{\partial S}{\partial t} \; . \tag{3.148}$$

Inserting (3.147) into (3.148), we obtain the fundamental equation in the Hamilton–Jacobi formulation of classical mechanics [5,108,109]:

$$\frac{1}{2m}\left(\frac{\partial S}{\partial x}\right)^2 + V(x,t) = -\frac{\partial S}{\partial t} \; , \tag{3.149}$$

which is a nonlinear partial differential equation.

3.5 Hamiltonianization of Physics

So far we have followed a stochastic approach to investigate the interaction between a Brownian particle and its environment. Here the concept of probability is essential because the system is open. The environmental influence takes place through the friction and diffusion coefficients present in the Langevin and Fokker–Planck equations. When the environmental influence via fluctuations is neglected, we obtain deterministic systems described by dissipative Newtonian equations. Insofar as the dissipative forces are not relevant either, the system is taken as isolated from its environment. This quite special class of Newtonian dynamical system can be set into a Hamiltonian framework.

We now wish to present an alternative way of describing a Brownian particle immersed in a fluid. This approach is based on the fundamental principle of Hamiltonianization of physics: in a system consisting of many particles, each particle constitutes a physical entity completely isolated from every other particle, in the sense that its motion is governed by a Hamiltonian function of the form

$$H(x,p) = \frac{p^2}{2m} + V(x) \; , \tag{3.150}$$

indicating that the total energy is a conserved quantity. A variational principle based on the action function [5,108,109]

$$S = \int_{t_1}^{t_2} (p\dot{x} - H)\mathrm{d}t \tag{3.151}$$

generates the Hamilton equations

$$\frac{\mathrm{d}p}{\mathrm{d}t} = -\frac{\partial H}{\partial x} \; , \quad \frac{\mathrm{d}x}{\mathrm{d}t} = \frac{\partial H}{\partial p} \; , \tag{3.152}$$

which are the Newtonian equations for a conservative system. In this sense the Newtonian equations of motion are derived from a Hamiltonian function through the principle of least action. From the deterministic initial conditions $q_0 = q(0)$ and $p_0 = p(0)$, the time evolution of the particle is exactly determined, i.e., with probability one, by solving the Hamilton equations (3.152). However, if for any physical reason the initial conditions cannot be specified with absolute accuracy, we are forced to use a probability density function $\mathcal{F}(p, x, t)$ whose time evolution is governed by the conservative Liouville equation:

$$\frac{\partial \mathcal{F}}{\partial t} = \{H, \mathcal{F}\} , \tag{3.153}$$

where $\{H, \mathcal{F}\}$ is the Poisson bracket of H and $\mathcal{F}$ defined by

$$\{H, \mathcal{F}\} = \frac{\partial H}{\partial x}\frac{\partial \mathcal{F}}{\partial p} - \frac{\partial H}{\partial p}\frac{\partial \mathcal{F}}{\partial x} . \tag{3.154}$$

According to this approach, Brownian motion is investigated through a Hamiltonian function containing all details of the interaction potential between the particles. In principle, this Brownian dynamics is elucidated by solving the Hamilton equations from deterministic initial conditions. In practice we can try to concretize this Hamiltonian dream by answering two questions:

- How can we derive stochastic differential equations from a Hamiltonian model?
- How can we describe dissipation and fluctuation phenomena within the Hamiltonian (or Lagrangian) framework, accounting for both the Brownian particle and the thermal reservoir as a completely isolated whole?

3.5.1 The Hamiltonian Dream

Even though the Langevin and Fokker–Planck equations provide correct results for Brownian motion, they suffer from the defect that they include the friction coefficient as a phenomenological parameter. The friction force is a consequence of (conservative) molecular forces and molecular structure in a way that is not described by the theory [70].

It is therefore important to set up a statistical mechanics of dissipative phenomena based on the following conditions:

- the dynamics of the particles has to be governed by the Hamilton equations,

- the initial state of the particles may be given by a probability density function.

Let us quote some objectives of this Hamiltonian statistical mechanics [34,70]:

- To justify the use of any statistical hypothesis on the basis of the molecular structure and molecular dynamics of matter in a such way that all physical properties are always determined with probability one (see [34,110,111]).
- To derive the friction and fluctuation coefficients from molecular forces between the particles (see [70,112]).
- To show how the Langevin and Fokker–Planck equations are to be obtained as a special case within a Hamiltonian framework, i.e., how to eliminate the environment variables. This is relevant to fixing the conditions of validity for these phenomenological equations of motion (see [34,70,111–125]).

The present status of the Hamiltonian program for the foundations of statistical mechanics may be appreciated by reading [34,110,126–128]. As a last comment we would just like to point out that it is also possible to elaborate a non-Hamiltonian form of statistical mechanics [129]:

> [...] the approach to true equilibrium is governed by interactions between the system and the outside world, not by interactions within the system itself. [...] Either we make a causal, Hamiltonian description of the whole Universe, or else we must allow for an essential element of randomness in the description of the motion of the limited system under study. [...] The random element here is not due to accidental shortcomings of the observer, but rather to the fact that the observer restricts his observations to a finite part of the Universe. [...] Statistical mechanics is not the mechanics of large, complicated systems; rather it is the mechanics of limited, not completely isolated systems.

3.5.2 Caldeira–Leggett Approach

Rather than establishing ab initio a complete Hamiltonian statistical mechanics from which the Langevin and Fokker–Planck equations are deduced as approximations, Caldeira and Leggett [130–132] introduced a different method, starting from the experimental validity of such phenomenological equations in order to construct a Hamiltonian model. We now present this approach.

Let us consider a Brownian particle described by the Langevin equations [see (3.121), (3.122) and (3.130)]

$$\frac{\mathrm{d}P}{\mathrm{d}t} = -\frac{\mathrm{d}V(X)}{\mathrm{d}X} - \alpha P + b_1\Phi(t) , \tag{3.155}$$

$$\frac{\mathrm{d}X}{\mathrm{d}t} = \frac{P}{m} , \tag{3.156}$$

with $\langle\Phi(t)\rangle = 0$ and $\langle\Phi(t_1)\Phi(t_2)\rangle = 2b_1\delta(t_1 - t_2)$. The phenomenological parameters m, $\alpha = \gamma/m$, and $b_1 = (\gamma m k_\mathrm{B} T)^{1/3}$ are in principle calculated experimentally. On the basis of these equations (3.155) and (3.156), we aim to set up a Hamiltonian model describing the Brownian motion [130–132].

Let us consider a system A with mass m (our Brownian particle) and Hamiltonian

$$H_A(x,p) = \frac{p^2}{2m} + V(x) . \tag{3.157}$$

Since any individual degree of freedom of the thermal reservoir (system B) is only weakly influenced by system A, we can suppose that it consists of an ensemble of harmonic oscillattors with mass M and frequency ω_n. Its Hamiltonian is given by

$$H_B(p_1, \ldots, p_N, q_1, \ldots q_N) = \sum_{n=1}^{N} \left(\frac{p_n^2}{2M} + \frac{M\omega_n^2}{2} q_n^2 \right) . \tag{3.158}$$

The justification for treating the environment as an oscillator reservoir lies in the simplicity of the model: the equations of motion are easily solved. However, one hopes to draw some general conclusions despite this restriction.

From Fig. 3.6, we see that the dissipation mechanism occurs precisely when we couple systems A and B, initially assumed to be isolated, using an interaction Hamiltonian $H_\mathrm{I}(p_1, \ldots, p_N, q_1, \ldots, q_N, p, x)$, so that the Hamiltonian of the total system has the form

$$H_\mathrm{T} = H_A + H_B + H_\mathrm{I} . \tag{3.159}$$

How can we construct the interaction Hamiltonian H_I? It is essential that such a term should not bring about unphysical results, i.e., the Hamiltonian model has to be constructed in order to reproduce the same physics as in the Langevin equations (3.155) and (3.156). In order to avoid the appearance of infinite renormalization potentials,

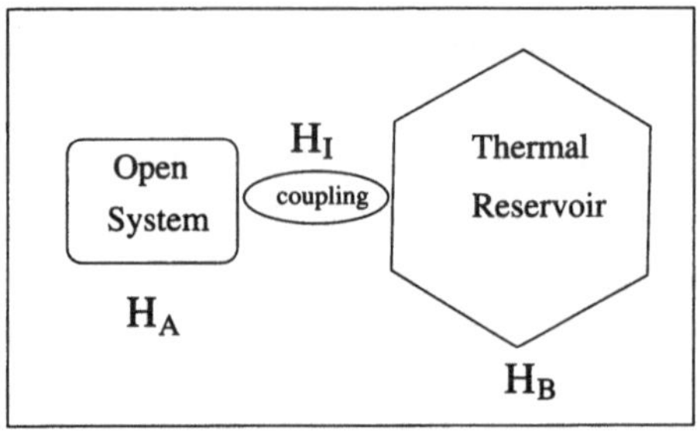

Fig. 3.6. In the Caldeira–Leggett approach, the open system is coupled to a thermal reservoir which is assumed to be made up of a set of harmonic oscillators

for example, we impose the following expression for the interaction Hamiltonian [133]:

$$H_{\mathrm{I}} = \sum_{n=1}^{N} \frac{M\omega_n^2}{2}(x^2 - 2xq_n) , \tag{3.160}$$

so that the (system A)–(system B) coupling is strictly linear. The total Hamiltonian (3.159) then becomes

$$H_{\mathrm{T}} = \frac{p^2}{2m} + V(x) + \sum_{n=1}^{N} \left[\frac{p_n^2}{2M} + \frac{M\omega_n^2}{2}(q_n - x)^2 \right] , \tag{3.161}$$

from which we obtain the oscillator dynamics

$$M\frac{\mathrm{d}^2 q_n}{\mathrm{d}t^2} + M\omega_n^2 q_n = M\omega_n^2 x , \tag{3.162}$$

and the particle dynamics

$$m\frac{\mathrm{d}^2 x}{\mathrm{d}t^2} + M\omega_n^2 x = M\omega_n^2 q_n - \frac{\partial V}{\partial x} . \tag{3.163}$$

Taking the solution of (3.162), i.e.,

$$\begin{aligned} q_n(t) = x(t) + (q_n' - x')\cos[\omega_n(t-t')] + \frac{p_n'}{M\omega_n}\sin[\omega_n(t-t')] \\ - \int_{t'}^{t} \cos[\omega_n(t-s)]\dot{x}(s)\mathrm{d}s , \end{aligned} \tag{3.164}$$

expressed in terms of $x(t)$ and the values of $q'_n = q_n(t')$, $p'_n = p_n(t')$ and $q'_n = q_n(t')$ at the initial time t', and substituting it into (3.163), we obtain an equation in terms of system A variables alone:

$$\frac{\mathrm{d}p}{\mathrm{d}t} + \int_{t'}^{t} \gamma(t-s)\frac{p(s)}{m}\mathrm{d}s + \frac{\partial V}{\partial x} = \Omega(t) \,, \tag{3.165}$$

$$\frac{\mathrm{d}x}{\mathrm{d}t} = \frac{p}{m} \,, \tag{3.166}$$

where the environmental properties appear thorough the friction kernel

$$\gamma(t-s) = \sum_{n=1}^{N} M\omega_n^2 \cos[\omega_n(t-s)] \tag{3.167}$$

and time-dependent force

$$\Omega(t) = \sum_{n=1}^{N} M\omega_n^2 \left\{ (q'_n - x')\cos[\omega_n(t-t')] + \frac{p'_n}{M\omega_n}\sin[\omega_n(t-t')] \right\} \,. \tag{3.168}$$

Notice that both (3.165) and (3.166) are still deterministic equations. Now we establish the following randomization procedure: we assume that we cannot specify the initial conditions of the particles of the thermal reservoir with absolute precision and that the corresponding probability density function is therefore given by

$$F(p'_n, q'_n) \propto \exp\left[-\frac{H(p'_n, q'_n - x')}{k_{\mathrm{B}}T} \right] \,, \tag{3.169}$$

where

$$H(p'_n, q'_n - x') = \frac{(p'_n)^2}{2M} + \frac{M\omega_n^2}{2}(q'_n - x)^2 \,. \tag{3.170}$$

Equation (3.169) yields

$$\langle q'_n \rangle = x' \,, \quad \langle p'_n \rangle = 0 \,, \quad \langle q'_n p'_n \rangle = 0 \,,$$
$$\langle (p'_n)^2 \rangle = Mk_{\mathrm{B}}T \,, \quad \langle (Q'_n - q')^2 \rangle = k_{\mathrm{B}}T/M\omega_n^2 \,,$$

so that the function $\Omega(t)$ turns out to have the statistical properties

$$\langle \Omega(t) \rangle = 0 \,, \tag{3.171}$$

$$\langle \Omega(t)\Omega(t') \rangle = k_{\mathrm{B}}T\gamma(t-t') \,. \tag{3.172}$$

Consequently, (3.165) and (3.166) become non-Markovian stochastic differential equations.

A Markovian approximation can be obtained by considering an infinite number of particles in the thermal reservoir in such a way that the frequencies ω_n can be continuously distributed. Taking this condition into account, one replaces (3.167) by the integral

$$\begin{aligned}\gamma(t-s) &= \int_{-\omega'}^{\omega'} \mathrm{d}\omega \rho(\omega) M\omega^2 \cos[\omega(t-s)] \\ &= \int_{-\omega'}^{\omega'} \mathrm{d}\omega \frac{2m\gamma}{\pi} \theta(\omega'-\omega) \cos[\omega(t-s)] \,, \end{aligned} \tag{3.173}$$

where

$$\rho(\omega) = \frac{2m\gamma}{\pi M\omega^2} \theta(\omega'-\omega) \tag{3.174}$$

is the frequency density, with $\theta(\omega'-\omega) = 1$ if $\omega' > \omega$, and $\theta(\omega'-\omega) = 0$ if $\omega' \leq 0$. Equation (3.172) can be approximated as

$$\gamma(t-s) \approx 2m\gamma\delta(t-s) \,. \tag{3.175}$$

This approximation is valid for $(\omega')^{-1} \ll (t-s)$ and is related to the interaction time between systems A and B. Inserting (3.174) into (3.165), we arrive at the Markovian Langevin equations (3.155) and (3.156) from which we started:

$$\frac{\mathrm{d}P}{\mathrm{d}t} = -\frac{\mathrm{d}V}{\mathrm{d}X} - \frac{\gamma}{m} P + \Omega(t) \,, \tag{3.176}$$

$$\frac{\mathrm{d}X}{\mathrm{d}t} = \frac{P}{m} \,, \tag{3.177}$$

with

$$\langle \Omega(t) \rangle = 0 \,, \tag{3.178}$$

$$\langle \Omega(t)\Omega(t') \rangle = 2m\gamma k_{\mathrm{B}} T \delta(t-t') \,. \tag{3.179}$$

Notice that the stochastic force $\Omega(t)$ generated by the Hamiltonian model is related to the true stochastic force in (3.155) by the relation

$$\Omega(t) = b_1 \Phi(t) \,. \tag{3.180}$$

Hence the Hamiltonian model [the function (3.161) and accompanying assumptions] is equivalent to the Langevin equations (3.155) and (3.156).

Summing up, the Caldeira–Leggett approach comprises two steps:

- experimental determination of the phenomenological coefficients present in the stochastic differential equations [e.g., the Langevin equations (3.155) and (3.156)],

- construction of a Hamiltonian model in which the parameters of the model are identified with the physical parameters of the stochastic differential equations [e.g., the relation (3.174)].

Because the Caldeira–Leggett approach starts from a phenomenological equation, it has proved quite fruitful in addressing the conceptual problems relating classical and quantum physics, but also in investigating practical applications in several contexts (e.g., quantum tunneling). This happens because Caldeira and Leggett make use of a quantization method based on Hamiltonian (or Lagrangian) functions:

$$\begin{array}{c}\text{Quantization}\\ \Uparrow\\ \text{Hamiltonian model}\\ \uparrow\\ \text{Langevin equations}\end{array}$$

thus confirming the common claim that the Hamiltonian formalism lies at the basis of quantum mechanics [109].

In the next chapter we shall investigate the problem of quantization of open systems, that is, the classical $\rightarrow$ quantum transition, and we shall show the relevance of setting up quantization methods directly from the equations of motion, independently of the formulation of a Hamiltonian model as required by Caldeira and Leggett:

$$\begin{array}{c}\text{Langevin equations and Fokker–Planck equation}\\ \Downarrow\\ \text{quantization}\end{array}$$

4 Quantum Physics

4.1 Motivation: The Problem of Quantization

In Chap. 3, we saw that a system of mass m interacting with its environment may be described by dissipative, nonlinear stochastic differential equations of the form

$$\frac{\mathrm{d}P}{\mathrm{d}t} = -\frac{\gamma}{m}P - \frac{\partial V}{\partial X} + b_1(t)\Phi(t) , \tag{4.1}$$

$$\frac{\mathrm{d}X}{\mathrm{d}t} = \frac{P}{m} + b_2(t)\Phi(t) . \tag{4.2}$$

Due to the statistical properties of the force $\Phi(t)$, the resulting stochastic process is in general non-Gaussian and non-Markovian. Insofar as the quantum nature of the environment is taken into account, the friction constant γ and both diffusion coefficients b_1 and b_2 turn out to depend on the Planck constant $\hbar$ (Planck's constant h divided by 2π). For example, in superconductor physics [131,134], the friction constant γ is interpreted as a resistance R in an electronic device, while Φ is treated as a stochastic current I, the capacitance C plays the role of the particle mass m, and the Brownian variable corresponding to the position X is the phase difference across a Josephson junction. In [134], we encounter the following physical situation: at low temperatures, the thermal reservoir is treated in a quantum theoretical way so that the diffusion coefficient exhibits some quantum effects wherein the stochastic force Φ has the following statistical properties:

$$b_2(t) = \int_0^t \langle \Phi(t') \rangle \mathrm{d}t' = 0$$

and

$$b_1(t) = \int_0^t \langle \Phi(t_1)\Phi(t_2) \rangle \mathrm{d}t_1 = \frac{1}{\pi}\int_0^t \int_{-\infty}^{\infty} \mathrm{e}^{-\mathrm{i}\nu\tau} S(\nu) \mathrm{d}\nu \mathrm{d}\tau ,$$

with $S(\nu)$ the spectral density expressed in terms of the Josephson frequency ν. The Brownian particle is still classical and is described by the Langevin equations in the form (4.1) and (4.2). From this model, one can measure the spectral density of the stochastic current $S_I(\nu)$ generated by the shunt resistance R. In Fig. 4.1, these experimental data are plotted for temperatures of 4.2 K (solid circles) and 1.6 K (open circles). The solid lines correspond to the theoretical result calculated using the spectral density

$$S(\nu) = \frac{4h\nu}{R}\left(\frac{1}{e^{h\nu/k_\mathrm{B}T}-1}+\frac{1}{2}\right) ,$$

which predicts zero-point fluctuations, while the dashed lines correspond to

$$S(\nu) = \frac{4h\nu}{R}\frac{1}{e^{h\nu/k_\mathrm{B}T}-1} ,$$

which predicts an absence of zero-point fluctuations.

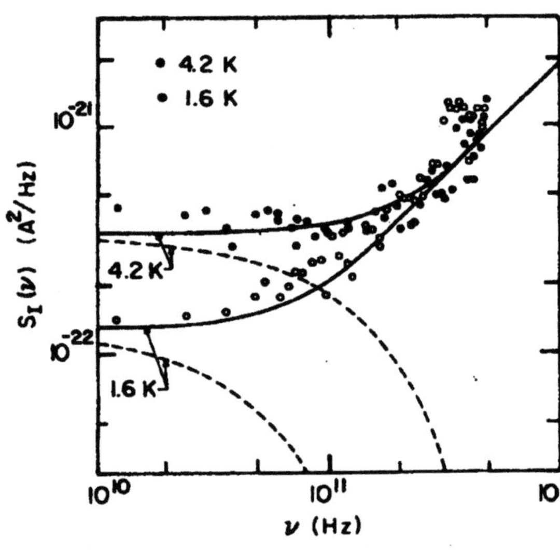

Fig. 4.1. Experimental data from superconductor physics plotted for temperatures of 4.2 K (*solid circles*) and 1.6 K (*open circles*). The meaning of the *solid* and *dashed curves* is explained in the text [134]

Although the approach used by Koch et al. accords with some experimental results [135], it has a very limited domain of validity because it is restricted by the classical nature of the Langevin equations

[134–136]. How then can we account for the underlying physics when the quantum nature of the Brownian particle is relevant? In other words, how can we quantize an open system?

In order to answer these questions, we must address the problem of quantization:

- firstly, we must bear in mind that every quantization process is a physical phenomenon,
- secondly, due to the physical character of quantization, we need to set up a mathematical procedure for describing it,
- finally, having established the physical and mathematical basis of the quantization process, we must investigate its conceptual or philosophical consequences for the elucidation of the relationship between classical and quantum physics. For example, is there any trajectory concept in quantum physics? Does the motion of a quantum particle describe a trajectory, like a planet or a Brownian particle?

With respect to such requirements, the present chapter aims to examine the plausibility of several quantization methods in the literature, such as those elaborated by Planck and Bohr (Sect. 4.2), Heisenberg (Sect. 4.3), Schrödinger (Sect. 4.4), Dirac (Sect. 4.5), Feynman (Sect. 4.6), Nelson (Sect. 4.7) and Olavo (Sect. 4.8). In Sect. 4.9, we present our own quantization scheme.

In Sect. 4.10, we criticize some attempts to establish quantum physics without a quantization process.

4.2 Planck–Bohr Quantization

The first few glimpses of the quantum world were caught by Planck in 1900 when he postulated that energy has to be understood as a discrete quantity in the emission and/or absorption of radiation within a blackbody [137]. In 1913, on the basis of Planck's hypothesis, Bohr also postulated that the angular momentum associated with electron motion around a nucleus should be quantized, i.e., that it should be an integer multiple of $\hbar$ [138]. Wilson [139], Planck [140], Sommerfeld [141], Schwarzschild [142], and Epstein [143] summed up the postulates of Bohr and Planck in the following ad hoc quantization rule [7,144]: given a Hamiltonian function $H(x,p)$ describing a bound physical system, some constants of motion I, e.g., the angular momentum, obey the condition (one degree of freedom)

$$I = \frac{1}{2\pi} \oint \mathrm{p}\mathrm{d}x = \nu\hbar \qquad (\nu = 0, 1, 2, \ldots) , \tag{4.3}$$

where the integral is taken along the trajectory which the system actually follows during a cycle of the motion of the coordinate x. The quantized values of the energy are obtained when $E = H(I)$. Assumption (4.3) is not mathematically well-defined, since it is not invariant under canonical transformations [145], and neither does it agree entirely with the experimental data: half-integer quantum numbers are not taken into account.

The basic feature of the quantization procedure developed by Planck, Bohr, Sommerfeld, and others is that it starts from the concept of classical trajectory and then calculates energy levels from there. Recently, this approach has permeated attempts to introduce chaos into quantum physics [146–148]. However, it is important to point out in this context that the trajectory concept in the quantum domain arises only through a formal analogy. Furthermore, it is difficult to imagine an extension of any quantization scheme on the basis of (4.3), or its generalizations, for open systems.

4.3 Heisenberg Quantization

Dissatisfied with the Bohr–Sommerfeld rule (4.3), Heisenberg [8,149] proposed a quantization method starting with the classical equations of motion and postulating that, in the quantum domain, both physical quantities momentum P and position X (c-numbers) should be replaced, respectively, by operators $\hat{P}$ and $\hat{X}$ (q-numbers), that is,

$$P \Longrightarrow \hat{P} , \qquad X \Longrightarrow \hat{X} . \tag{4.4}$$

These operators obey an algebraic structure based on the Born–Jordan–Dirac relations [10,150]

$$[\hat{X}, \hat{X}] = 0 = [\hat{P}, \hat{P}] , \tag{4.5}$$

$$[\hat{X}, \hat{P}] = \mathrm{i}\hbar , \tag{4.6}$$

where $[\hat{A}, \hat{B}] = \hat{A}\hat{B} - \hat{B}\hat{A}$ is the so-called commutator of $\hat{A}$ and $\hat{B}$.

Employing the Heisenberg quantization (4.4–4.6) in the classical equations (4.1) and (4.2), we obtain

$$\frac{\mathrm{d}\hat{P}}{\mathrm{d}t} = -\frac{\gamma}{m}\hat{P} - \frac{\partial}{\partial \hat{X}} V(\hat{X}) + b_1(t)\hat{\Phi}(t) , \tag{4.7}$$

$$\frac{\mathrm{d}\hat{X}}{\mathrm{d}t} = \frac{\hat{P}}{m} + b_2(t)\hat{\Phi}(t) . \tag{4.8}$$

The friction constant γ, the mass m and both diffusion coefficients b_1 and b_2 are c-numbers, whereas the quantum stochastic force $\hat{\Phi}(t)$ satisfies some algebraic constraints, such as

$$[\hat{\Phi}(t), \hat{\Phi}(t')] = \mathcal{R}(t, t') , \tag{4.9}$$

$$\langle[\hat{\Phi}(t), \hat{\Phi}(t')]_+\rangle = \mathcal{S}(t, t') , \tag{4.10}$$

where $[\hat{A}, \hat{B}]_+ = \hat{A}\hat{B} + \hat{B}\hat{A}$ denotes the anticommutator of $\hat{A}$ and $\hat{B}$, and $\mathcal{R}$ and $\mathcal{S}$ are functions (c-numbers) containing all the physical properties that characterize the environment (e.g., friction coefficient γ, Planck's constant $\hbar$, Boltzmann's constant k_{B}, and the absolute temperature T of the thermal reservoir). Equations (4.7) and (4.8) obeying (4.9) and (4.10) are called the quantum Langevin equations [116,136,151–154].

Heisenberg's quantization method has the advantage of being based directly on the equations of motion without needing to formulate Hamiltonian models [151]. However, as a consequence of the changed algebraic structure, moving from a c-number theory to a q-number theory, the concept of trajectory disappears from quantum physics. Besides this conceptual problem, another (theoretical) difficulty inherent in this quantization scheme is related to the ambiguities arising from the operator ordering in the calculation of physical quantities [116,155,156]:

> Even when the interaction of an atom with its radiation field is treated fully quantum mechanically, the use of the Heisenberg picture and so-called 'quantum Langevin equations' may lead to interpretation problems which partially have their origin in the classical pictures used. The interpretation of single terms in the equations often depends on the operator ordering. For example, in normal ordering spontaneous emission is connected with the radiation reaction while the Langevin force is associated with vacuum fluctuations. When a symmetric ordering (motivated by the requirement of Hermitian Heisenberg operators in the quantum Langevin equation) is employed instead, radiation reaction and vacuum fluctuations give equal contributions to spontaneous emission. Such discussions demonstrate that the Heisenberg picture is inappropriate for the interpretation of quantum states and quantum processes.
>
> Quoted from [157]

4.4 Schrödinger Quantization

In order to reveal the true nature of the quantum conditions of the Planck–Bohr–Sommerfeld approach (4.3), Schrödinger [9] proposed a different method of quantization: one starts with the Hamilton–Jacobi equation (3.149) for the action function $S(x,t)$ and one introduces the relation

$$S(x,t) = -\mathrm{i}\ell \ln \psi(x,t) , \tag{4.11}$$

where ℓ has the dimensions of action and ψ is a complex-valued function. Inserting (4.11) into the Hamilton–Jacobi equation (3.149), one builds a functional $\mathcal{F}(\psi, \psi^\dagger)$ and assumes that the integral

$$\mathcal{I}(\psi, \psi^\dagger) = \int_{-\infty}^{\infty} \psi\psi^\dagger \mathrm{d}x \tag{4.12}$$

is an extremum $\delta\mathcal{I} = 0$. Applying this variational principle and taking the quantum limit

$$\ell \to \hbar , \tag{4.13}$$

one arrives at the Schrödinger equation describing an isolated system, viz.,

$$\mathrm{i}\hbar \frac{\partial}{\partial t}\psi(x,t) + \frac{\hbar^2}{2m}\frac{\partial^2}{\partial x^2}\psi(x,t) - V(x,t)\psi(x,t) = 0 . \tag{4.14}$$

Such a quantization method is used, for example, by Schönberg [158] for nonlinear generalizations of the Schrödinger and Dirac equations, as well as for deriving dissipative quantum equations of motion [159]. Despite these extensions, the Schrödinger quantization procedure was severely criticized by Schrödinger himself [160]:

- The relation between S and ψ given by (4.11) is incomprehensible (*unverständlich*). Indeed, Schrödinger had originally considered $S = \ell \ln \psi$, with a real function ψ.
- The formulation of the integral (4.12) is not entirely devoid of ambiguities (*nicht ganz eindeutig*).
- Finally, the very existence of the variational principle ($\delta\mathcal{I} = 0$) is incomprehensible (*unverständlich*). (Frieden has attempted to make the Schrödinger quantization procedure comprehensible on the basis of the Fisher information [161].)

To get around these difficulties, Schrödinger dropped his quantization method and tried to establish a more secure foundation for his equation by making use of the Hamilton analogy between mechanics and optics [7,160]. However, it is worth noticing that neither procedure constitutes a logical derivation of the Schrödinger equation from deeper physical principles. This equation has to be taken as a fundamental postulate of quantum mechanics. What matters is the agreement between its consequences and the results of experiments [162]. Hence, any search for a suitable procedure of quantization, i.e., a method for obtaining quantum mechanical equations of motion from the equations of classical mechanics, turns out to be considered as a mere pseudoproblem. Further consideration of the problem of setting up quantum physics without a quantization process can be found in Sect. 4.10.

4.5 Dirac Quantization

According to Dirac [10,163], there is a close connection between the Hamiltonian formalism of classical mechanics and quantum mechanics. We must start with a Hamiltonian function and then use the Heisenberg quantization in the coordinate representation in order to arrive at the Schrödinger equation. Dirac quantization or canonical quantization works as follows:

- An isolated classical system is described by a time-independent Hamiltonian function $H(p,x)$ identified with the total energy of the system, viz.,

$$H(p,x) = \frac{p^2}{2m} + V(x) = E \,. \tag{4.15}$$

- Having imposed the criterion (4.15), we use the canonical quantization rules

$$p \Longrightarrow \hat{p} = -\mathrm{i}\hbar\frac{\partial}{\partial x} \,, \tag{4.16}$$

$$x \Longrightarrow \hat{x} = x \,, \tag{4.17}$$

$$E \Longrightarrow \hat{E} = \mathrm{i}\hbar\frac{\partial}{\partial t} \,, \tag{4.18}$$

  and thereby obtain the Schrödinger equation

$$\hat{H}\psi = \mathrm{i}\hbar\frac{\partial}{\partial t}\psi \,, \tag{4.19}$$

where $\psi = \psi(x,t)$ is a complex function. Algebraically, Dirac quantization can be formulated by replacing the Poisson brackets by the corresponding commutators [10,149,163–166]:

$$\{A,B\} = \frac{\partial A}{\partial x} - \frac{\partial B}{\partial p} \quad \Longrightarrow \quad \frac{[\hat{A},\hat{B}]}{\mathrm{i}\hbar} = \frac{\hat{A}\hat{B} - \hat{B}\hat{A}}{\mathrm{i}\hbar} , \tag{4.20}$$

or in terms of the Liouville equation

$$\frac{\partial \rho}{\partial t} = \{H,\rho\} \quad \Longrightarrow \quad \frac{\partial \hat{\rho}}{\partial t} = \frac{\mathrm{i}}{\hbar}[\hat{H},\hat{\rho}] , \tag{4.21}$$

which is the von Neumann equation for the density operator $\hat{\rho}$.

Even though the Dirac procedure correctly reproduces the equations of motion of quantum mechanics, we should point out some drawbacks regarding this quantization method:

- It is a quantization scheme privileging the Cartesian coordinates [7,167] (see Appendix C).
- As remarked by Dirac himself [165], this quantization scheme is not mathematically well-defined due to the problem of operator ordering in the transition from c-numbers to q-numbers [168,169].
- The wave function ψ upon which the operators $\hat{x}$ and $\hat{p}$ act arises ex nihilo in the theory, so no light is thrown on its physical interpretation.
- As a consequence of this quantization procedure, no trajectory concept arises in the quantum domain.

In spite of the difficulties quoted above, it is still possible to use the Dirac rules to quantize Brownian motion. Hamiltonian models can be elaborated where both Brownian particle and thermal bath can be imagined as forming an isolated whole described by a Hamiltonian of the form [116,130,132,152–154,170–172]

$$H = H_A + H_B + H_{\mathrm{I}} . \tag{4.22}$$

In particular, the most general form of Hamiltonian compatible with the stochastic Langevin equations and with the harmonic constitution of the heat bath (see Sect. 3.5.2) is given by the Caldeira–Leggett Hamiltonian [132]

$$\begin{aligned} H(x,p;q_n,p_n) = \frac{p^2}{2m} + V(x) + \sum_{n=1}^{N}\left(\frac{p_n^2}{2M} + \frac{M}{2}\omega_n^2 x_n^2\right) \\ - \sum_{n=1}^{N}\Big[R_n(x,p)q_n + S_n(p,x)p_n\Big] + F(p,x) , \end{aligned} \tag{4.23}$$

which involves an ordering problem when quantized via the Dirac procedure. In order to circumvent this difficulty, one restricts the interaction Hamiltonian to the case of linear coupling (3.160). According to this approach:

- Caldeira and Leggett [130,171] quantize a Hamiltonian of the form (4.23) and use the Feynman–Vernon functional theory [173] in terms of a density matrix.
- As the system as a whole described by the Hamiltonian (4.23) is isolated, Cavalcanti et al. [174–178] investigate the Schrödinger function associated with the Brownian particle. This approach is still restricted to linear cases.
- Toda [113] quantizes a Hamiltonian of the form (4.23) by replacing Poisson brackets with commutators (4.21) using perturbation theory in terms of a coupling constant.
- Finally, Ford et al. [116,152–154,172] derive the quantum Langevin equations from a Hamiltonian model based on (4.23).

4.6 Feynman Quantization

Rather than seeking Hamiltonian functions, Feynman [11] followed Dirac's suggestions [179] and proposed a method of quantizing a given isolated system initially described by a classical Lagrangian function of the form

$$L(x, \dot{x}) = \frac{m}{2}\dot{x}^2 - V(x) \, . \tag{4.24}$$

Let $\Omega(x,t)$ be a function connected at points $(x_\mathrm{i}, t_\mathrm{i})$ and $(x_\mathrm{f}, t_\mathrm{f})$ by the integral equation

$$\Omega(x_\mathrm{f}, t_\mathrm{f}) = \int_{-\infty}^{\infty} \Pi(x_\mathrm{f}, t_\mathrm{f}; x_\mathrm{i}, t_\mathrm{i}) \Omega(x_\mathrm{i}, t_\mathrm{i}) \mathrm{d}x_\mathrm{i} \, , \tag{4.25}$$

where $t_\mathrm{f} > t_\mathrm{i}$. We assume that the propagator $\Pi(x_\mathrm{f}, t_\mathrm{f}; x_\mathrm{i}, t_\mathrm{i})$ is given by

$$\Pi(b, a) = \int_a^b \exp\left[\frac{\mathrm{i}}{\hbar}\int_{t_\mathrm{i}}^{t_\mathrm{f}} L(x, \dot{x}, t)\mathrm{d}t\right] \mathcal{D}x(t) \, , \tag{4.26}$$

where we have set $a = (x_\mathrm{i}, t_\mathrm{i})$ and $b = (x_\mathrm{f}, t_\mathrm{f})$. The symbol $\mathcal{D}x(t)$ denotes a certain integration measure giving meaning to the sum over all possible paths $x(t)$ from $a = (x_\mathrm{i}, t_\mathrm{i})$ to $b = (x_\mathrm{f}, t_\mathrm{f})$ of the functional

$$\exp\left[\frac{\mathrm{i}}{\hbar}\int_{t_\mathrm{i}}^{t_\mathrm{f}} L(x, \dot{x}, t)\mathrm{d}t\right] \, , \tag{4.27}$$

which contains the Planck constant $\hbar$. To obtain the Schrödinger equation from (4.25), we must start with the time-independent Lagrangian (4.24) and discretize the time t into N intervals of width ε/N, such that $t_1 = t_\mathrm{i}$, $t_\mathrm{f} = t_N$ and $\varepsilon = t_N - t_1$. By the standard mean value theorem [180], the action

$$S(t_\mathrm{f}, t_\mathrm{i}) = \int_{t_\mathrm{i}}^{t_\mathrm{f}} L(x, \dot{x})\mathrm{d}t \tag{4.28}$$

can be replaced by

$$S(t_N, t_1) = \varepsilon \sum_{k=1}^{N} \alpha_k L\Big(x(t_k), \dot{x}(t_k)\Big) . \tag{4.29}$$

The dimensionless parameters α_k are needed to guarantee the differentiability of the trajectories. As a consequence of this discretization procedure, the action (4.29) calculated with the Lagrangian (4.24) becomes

$$S(t_N, t_1) = \varepsilon \sum_{k=1}^{N} \alpha_k \left[\frac{m}{2} \left(\frac{\gamma_k \eta}{\varepsilon} \right)^2 - V(x + b_k \eta) \right] , \tag{4.30}$$

where

$$\frac{\gamma_k \eta}{\varepsilon} = \frac{x(t_k) - x(t_{k-1})}{\varepsilon} , \quad b_k \eta = x(t_k) - x ,$$

$$b_k = \sum_{m=2}^{k} \gamma_m , \quad b_1 = 0 , \quad b_N = -1 .$$

The parameters b_k and γ_k are arbitrary and responsible for the smoothness of the accompanying function describing all possible paths (η being restricted to real values). Inserting (4.30) into (4.26) and choosing a determined integration measure $\mathcal{D}(t)$, we arrive at the following expression for (4.25) [180]:

$$\Omega(x, t + \varepsilon) = \frac{1}{\sqrt{\pi R}} \int_{-\infty}^{\infty} \mathrm{e}^{(-R\eta^2 + W)} \Omega(x + \eta, t) \mathrm{d}\eta , \tag{4.31}$$

where

$$R = -\frac{\mathrm{i}m\lambda}{2\hbar\varepsilon} , \quad \lambda = \sum_{k=1}^{N} \alpha_k \gamma_k^2 , \quad W = \frac{\mathrm{i}\varepsilon}{\hbar} \sum_{k=1}^{N} \alpha_k V(x + b_k \eta) .$$

Following Feynman [11,181], we expand $\exp[W(\eta)]$ and $\Omega(x+\eta,t)$ in (4.31) and integrate. We then obtain [180]

$$\Omega(x,t+\varepsilon)=\sum_{n=0}^{\infty}F_n^j(j_1,\ldots,j_s|l_1,\ldots,l_s)Q^j(j_1,\ldots,j_s|l_1,\ldots,l_s)\varepsilon^n\,, \tag{4.32}$$

summing over all non-negative integers $s,j,j_1,\ldots,j_s,l_1,\ldots,l_s$ compatible with the conditions

$$2n=j+\sum_{r=1}^{s}(2+j_r)l_r\,,$$

$$j_1>\ldots>j_s\,,\quad r\leq s\,,\quad l_r=0\Rightarrow s=1\,.$$

In (4.32), we have

$$Q^j(j_1,\ldots,j_s|l_1,\ldots,l_s)=\frac{\partial^j\Omega}{\partial x^j}\prod_{r=1}^{s}\left(\frac{\partial^{j_r}V}{\partial x^{j_r}}\right)^{l_r} \tag{4.33}$$

and

$$F_n^j(j_1,\ldots,j_s|l_1,\ldots,l_s)=-\frac{\mathrm{i}^l}{\hbar^l}\frac{J(d)}{j!l_1!\ldots l_s!}\prod_{r=1}^{s}\left(\frac{\Delta_{j_r}^{l_r}}{j_r!}\right)\,, \tag{4.34}$$

where

$$l=\sum_{r=1}^{s}l_r\,,\quad d=j+\sum_{r+1}^{s}k_jl_j\,,$$

and

$$J(d)=|d-1|!!\left(\frac{\mathrm{i}\hbar}{m\lambda}\right)^{d/2}\,.$$

Now we expand the function $\Omega(x,t+\varepsilon)$ in the form

$$\Omega(x,t+\varepsilon)=\sum_{n=1}^{\infty}\frac{\varepsilon^n}{n!}\frac{\partial^n}{\partial t^n}\Omega(x,t)\,, \tag{4.35}$$

and substitute it into (4.32). We thus obtain a hierarchy of equations by comparing powers of ε on both sides of (4.32):

$$\left[F_0^j(j_1,\ldots,j_s|l_1,\ldots,l_s)Q^j(j_1,\ldots,j_s|l_1,\ldots,l_s)\right]\varepsilon^0=0\,, \tag{4.36}$$

$$\left[\frac{\partial}{\partial t}\Omega(x,t) - F_1^j(j_1,\ldots,j_s|l_1,\ldots,l_s)Q^j(j_1,\ldots,j_s|l_1,\ldots,l_s)\right]\varepsilon^1 = 0\,, \tag{4.37}$$

$$\vdots$$

$$\left[\frac{1}{n!}\frac{\partial^n}{\partial t^n}\Omega(x,t) - F_n^j(j_1,\ldots,j_s|l_1,\ldots,l_s)Q^j(j_1,\ldots,j_s|l_1,\ldots,l_s)\right]\varepsilon^n = 0\,, \tag{4.38}$$

and so on.

The integral equation (4.31) is therefore equivalent to an infinite hierarchy of partial differential equations like (4.37) and (4.38). Feynman analyses this equivalence by restricting to the first order in ε. Then one obtains a trivial identity from (4.36) and one gets the Schrödinger equation for $\Omega(x,t) = \psi(x,t)$ from (4.37).

Why do we neglect terms of order ε^2 in the discretization procedure? In order to save his quantization procedure, Feynman [11] provides an operational interpretation of the time interval ε:

> Assume that we have a particle which can take up various values of a coordinate x. Imagine that we make an enormous number of successive position measurements, let us say separated by a small time interval ε. Then a succession of measurements such as $A, B, C, \ldots$, might be the succession of measurements of the coordinate x at successive times $t_1, t_2, t_3, \ldots$, where $t_{i+1} = t_i + \varepsilon$. Let the value, which might result from measurement of the coordinate at time t_i, be x_i. [...] From a classical point of view, the successive values $x_1, x_2, x_3, \ldots$, of the coordinate practically define a path $x(t)$. Eventually, we expect to go to the limit $\varepsilon \to 0$.

Although it works as a good computational tool [182,183], provided one controls the errors inherent in replacing an integral equation by discretized equations [184–187], the Feynman formalism suffers from serious drawbacks that make it unsuitable as a quantization process [180,188]:

- Mathematical difficulties arise in defining the integration measure in the Feynman integral in more general cases [11,180,189].
- There is some arbitrariness in the choice of the propagator or kernel Π in (4.25). The Feynman method does not provide the correct quantization of a double pendulum [190], for instance.
- Another source of ambiguity lies in the choice of the point at which the propagator (4.26) is evaluated for small time differences

ε [11,182,190,191]. In the Feynman method, this midpoint question corresponds to the operator ordering problem in the Dirac and Heisenberg quantizations [186,187,192–197].

- Feynman's assumption that for large times

$$\psi(x_\mathrm{f}, t_\mathrm{f}) = \int_{-\infty}^{\infty} \Pi(x_\mathrm{f}, t_\mathrm{f}; x_\mathrm{i}, t_\mathrm{i})\psi(x_\mathrm{i}, t_\mathrm{i})\mathrm{d}x_\mathrm{i} \ , \qquad (4.39)$$

where

$$\Pi(b; a) = \text{smooth function} \times \exp\left[\frac{\mathrm{i}}{\hbar} S_\mathrm{cl}(a, b)\right] \ , \qquad (4.40)$$

is only valid for a free particle and a harmonic oscillator [11, 173,180]. Therefore, the interpretation of the Schrödinger function $\psi(x, t)$ as being a kind of superposition of all possible classical trajectories of the system cannot be achieved in a general way.

Controlling the ambiguities and inconsistencies, Caldeira and Leggett [130] have shown that it is possible to quantize a Brownian particle by combining the Dirac or canonical rules with Feynman path integrals. Thus, associating a Lagrangian function with the Hamiltonian model presented in Sect. 3.5.2, using the Feynman–Vernon techniques based on the influence functional for the reduced density matrix [173]

$$\rho(x_1, x_2, t + \varepsilon) = \int_{-\infty}^{\infty}\int_{-\infty}^{\infty} J(x_1, x_2, t + \varepsilon; x_1', x_2', t)\mathrm{d}x_1'\mathrm{d}x_2' \ , \qquad (4.41)$$

and after a good deal of algebra, Caldeira and Leggett [130] derived the partial differential equation

$$\mathrm{i}\hbar\frac{\partial}{\partial t}\rho + \frac{\hbar^2}{2m}\left(\frac{\partial^2}{\partial x_1^2} - \frac{\partial^2}{\partial x_2^2}\right)\rho - [V_\mathrm{R}(x_1) - V_\mathrm{R}(x_2)] = \mathcal{O}\rho \ , \qquad (4.42)$$

tantamount to (4.41) to first order in ε. In (4.42), the term $\mathcal{O}\rho$ is given by

$$\mathcal{O}\rho = -\mathrm{i}\hbar\gamma(x_1 - x_2)\left(\frac{\partial}{\partial x_1} - \frac{\partial}{\partial x_2}\right)\rho - \frac{\mathrm{i}2m\gamma k_\mathrm{B}T}{\hbar}(x_1 - x_2)^2\rho \ . \qquad (4.43)$$

Equation (4.42) is called the Caldeira–Leggett master equation. V_R stands for a renormalized potential arising as a consequence of the assumed Hamiltonian model.

To close this section, we wish to emphasize that the ambiguity and arbitrariness present in the Heisenberg, Dirac and Feynman schemes are an artifact of quantization methods based on Lagrangians and/or Hamiltonians and do not reflect the underlying physics or the true relationship between classical and quantum physics. Thus, just as happens with the Heisenberg and Dirac quantization methods, we conclude together with Joos [157] that Feynman quantization is also "inappropriate for the interpretation of quantum states and quantum processes." Despite this fact it is important to take into account its heuristic value. Only in this sense could we understand the significance of the Schrödinger equation and the Caldeira–Leggett equation (4.42) obtained from this quantization method. [Equation (4.42) is considered further in Sect. 4.9.]

4.7 Nelson Quantization

In empty space a particle is assumed to undergo no frictional force and the action of a quantum environment whose diffusivity is given by [198]

$$\mathcal{D} = \frac{\hbar}{2m} . \tag{4.44}$$

Here only the position of the particle is to be considered as a stochastic process described by the equation

$$\frac{\mathrm{d}X}{\mathrm{d}t} = A(X,t) + \left(\frac{\hbar}{2m}\right)^{1/3} \Phi(t) , \tag{4.45}$$

with

$$\langle \Phi(t) \rangle = 0 \tag{4.46}$$

and

$$\langle \Phi(t_1)\Phi(t_2) \rangle = 2\left(\frac{\hbar}{2m}\right)^{1/3} \delta(t_1 - t_2) . \tag{4.47}$$

The time evolution of the probability density function $D_X(x,t)$ is governed by the Kolmogorov stochastic equation (2.20) or its adjoint. Using the arguments that lead to the Fokker–Planck equation (3.6), we arrive at

$$\frac{\partial D_X}{\partial t} = -\frac{\partial}{\partial x}\Big[B(x,t)D_X\Big] + \frac{\hbar}{2m}\frac{\partial^2}{\partial x^2}D_X \tag{4.48}$$

and its adjoint

$$\frac{\partial D_X}{\partial t} = -\frac{\partial}{\partial x}\Big[B^*(x,t)D_X\Big] - \frac{\hbar}{2m}\frac{\partial^2}{\partial x^2}D_X , \tag{4.49}$$

where the coefficients are calculated from

$$B(x,t) = \lim_{\epsilon \to 0} \frac{\langle X(t+\epsilon) - X(t) \rangle}{\epsilon} , \tag{4.50}$$

$$B^*(x,t) = \lim_{\epsilon \to 0} \frac{\langle X(t) - X(t-\epsilon) \rangle}{\epsilon} . \tag{4.51}$$

Adding (4.48) and (4.49), we obtain

$$\frac{\partial D_X}{\partial t} = -\frac{\partial}{\partial x}\Big[G(x,t) D_X\Big] , \tag{4.52}$$

where

$$G = \frac{1}{2}(B + B^*) . \tag{4.53}$$

On the other hand, subtracting (4.48) from (4.49), we have

$$\frac{\partial}{\partial x}\Big[U(x,t) D_X\Big] - \frac{\hbar}{2m}\frac{\partial^2}{\partial x^2} D_X = 0 , \tag{4.54}$$

where

$$U = \frac{1}{2}(B - B^*) . \tag{4.55}$$

Supposing

$$U = \frac{\hbar}{m}\frac{\partial}{\partial x} S , \qquad G = \frac{\hbar}{m}\frac{\partial}{\partial x} R ,$$

from the coupled equations (4.52) and (4.54), Nelson [198] obtains nonlinear equations involving the functions U and G. Then introducing

$$\psi = \mathrm{e}^{R+\mathrm{i}S} , \qquad D_X = |\psi|^2 ,$$

the Schrödinger equation (4.14) is derived, provided that

$$\frac{1}{2}(BB^* + B^*B) = -\frac{1}{m}\frac{\partial}{\partial x} V(x) . \tag{4.56}$$

According to Nelson, the Schrödinger equation is interpreted as describing a Markovian quantum particle without friction and having an actual trajectory defined by the stochastic differential equation (4.45).

Some objections against Nelson quantization are the following:

- The nonlinear partial differential equations coupling U and G derived from (4.52) and (4.54) are not always equivalent to the Schrödinger equation. Thus, the Schrödinger function cannot be seen as a mere mathematical consequence of (4.52) and (4.54) [199].

- Mathematically, the so-called forward and backward Kolmogorov stochastic equations (4.48) and (4.49) are on an equal footing. However, the physical meaning of the backward equation (4.49) is doubtful [200,201]. Thus, assumption (4.56) is considered by Nelson himself as a "somewhat arbitrary booking form" [202]. Recently, Olavo [203] has attempted to provide a physical justification of Nelson's assumption (4.56).
- The Schrödinger equation cannot be considered as equivalent to a classical Markovian stochastic process [200,201].

A detailed and updated study of applications, extensions and limitations of Nelson quantization can be found in [204].

4.8 Olavo Quantization

We turn now to a quantization method based on the entropy concept developed recently by Olavo [205]. We start from the Newtonian equations (3.136) and (3.137) for an isolated particle, generating the conservative Liouville equation

$$\frac{\partial}{\partial t} D_{XP}(x,p,t) + \frac{p}{m}\frac{\partial}{\partial x} D_{XP}(x,p,t) - \frac{\partial V(x)}{\partial x}\frac{\partial}{\partial p} D_{XP}(x,p,t) = 0 \,. \tag{4.57}$$

Multiplying (4.57) by p^n and integrate over p, we obtain

$$\frac{\partial}{\partial t}\mathcal{I}^{(n)}(x,t) - \frac{n}{m}\frac{\partial V(x)}{\partial x}\mathcal{I}^{(n-1)}(x,t) + \frac{\partial}{\partial x}\mathcal{I}^{(n+1)}(x,t) = 0 \,, \tag{4.58}$$

where we have defined $\mathcal{I}^{(n)}(x,t)$ by

$$\mathcal{I}^{(n)}(x,t) = \frac{1}{m^n}\int_{-\infty}^{\infty} p^n D_{XP}(x,p,t)\mathrm{d}p \,. \tag{4.59}$$

The function $\mathcal{I}^{(0)}(x,t)$ is the so-called marginal probability density

$$\mathcal{I}^{(0)}(x,t) = D_X(x,t) = \int_{-\infty}^{\infty} D_{XP}(x,p,t)\mathrm{d}p \,. \tag{4.60}$$

The Liouville equation (4.57) is therefore mathematically equivalent to the infinite set of coupled equations of motion (4.58).

Let us suppose that physically we may restrict (4.58) to the cases $n = 0$ and $n = 1$. Then for $n = 0$, (4.58) is written as a continuity equation

$$\frac{\partial}{\partial t}D_X(x,t)+\frac{\partial}{\partial x}\mathcal{I}^{(1)}(x,t)=0\,, \tag{4.61}$$

while for $n=1$, we have the equation

$$\frac{\partial}{\partial t}\mathcal{I}^{(1)}(x,t)-\frac{1}{m}\frac{\partial V(x)}{\partial x}D_X(x,t)+\frac{\partial}{\partial x}\mathcal{I}^{(2)}(x,t)=0\,. \tag{4.62}$$

We define the mean momentum $\pi(x,t)$ by

$$\pi(x,t)=\frac{\displaystyle\int_{-\infty}^{\infty}pD_{XP}(x,p,t)\mathrm{d}p}{\displaystyle\int_{-\infty}^{\infty}D_{XP}(x,p,t)\mathrm{d}p}=m\frac{\mathcal{I}^{(1)}}{\mathcal{I}^{(0)}} \tag{4.63}$$

and the fluctuation of p around $\pi(x,t)$ by

$$\sigma_p(x,t)=\frac{\displaystyle\int_{-\infty}^{\infty}(p-\pi)^2D_{XP}(x,p,t)\mathrm{d}p}{\displaystyle\int_{-\infty}^{\infty}D_{XP}(x,p,t)\mathrm{d}p}\,. \tag{4.64}$$

Then substituting (4.61) into (4.62), we arrive at [205]

$$\frac{1}{m}\frac{\partial}{\partial x}\sigma_p+\left(\frac{\partial\pi}{\partial t}+\frac{1}{2m}\frac{\partial\pi^2}{\partial x}+\frac{\partial V}{\partial x}\right)D_X(x,t)=0\,. \tag{4.65}$$

But is there any physical reason for restricting the infinite set of equations (4.58) to the equations (4.61) and (4.65)? According to Olavo [205], such a restriction is dictated by the quantum nature of matter. Olavo then formulates his quantization method as follows.

We must investigate the thermodynamic fluctuations δx of the position x. With this goal in mind, we consider a thermodynamically isolated system whose entropy is given by the Boltzmann principle:

$$\mathcal{S}(x,t)=k_\mathrm{B}\ln\Omega(x,t)\quad\text{or}\quad\Omega(x,t)=\mathrm{e}^{\mathcal{S}(x,t)/k_\mathrm{B}}\,, \tag{4.66}$$

where $\Omega(x,t)$ represents the accessible states when the position varies between x and $x+\delta x$. We expand the entropy function $\mathcal{S}(x+\delta x)$ for small displacements δx, taking into account the fact it must be a maximum. We then obtain

$$\mathcal{S}(x+\delta x)=\mathcal{S}_0(x)+\frac{1}{2}\frac{\partial^2\mathcal{S}_0}{\partial x^2}(\delta x)^2\,, \tag{4.67}$$

where $\mathcal{S}_0(x)$ is the equilibrium entropy. Inserting this result (4.67) into (4.66), we obtain

$$\Omega'(x, \delta x, t) = \Omega_0 \mathrm{e}^{-a(\delta x)^2} , \tag{4.68}$$

with

$$a = \frac{1}{2k_\mathrm{B}} \frac{\partial^2 S_0}{\partial x^2} (\delta x)^2 > 0 . \tag{4.69}$$

Supposing

$$D_X(x, t) \propto \Omega'(x, \delta x, t) , \tag{4.70}$$

we can calculate the fluctuation in x as

$$\sigma_x(x, t) = \frac{\displaystyle\int_{-\infty}^{\infty} (\delta x)^2 \mathrm{e}^{-a(\delta x)^2} \mathrm{d}(\delta x)}{\displaystyle\int_{-\infty}^{\infty} \mathrm{e}^{-a(\delta x)^2} \mathrm{d}(\delta x)} = \frac{1}{2a} . \tag{4.71}$$

We now impose a further condition to be obeyed by the equilibrium fluctuations in the position and momentum (4.64) and (4.71):

$$\sigma_x \sigma_p = \frac{\hbar^2}{4} . \tag{4.72}$$

Consequently, using (4.70), we find

$$\sigma_p = -\frac{\hbar^2}{4} \frac{\partial^2}{\partial x^2} \ln D_X(x, t) . \tag{4.73}$$

Substituting (4.73) into (4.65), considering $D_X(x, t) = R^2(x, t)$ and

$$\pi(x, t) = \frac{\partial S}{\partial x} , \tag{4.74}$$

we arrive at

$$R^2 \frac{\partial}{\partial x} \left[\frac{\partial S}{\partial t} + \frac{1}{2m} \left(\frac{\partial S}{\partial x} \right)^2 + V(x) - \frac{\hbar^2}{2mR} \frac{\partial^2 R}{\partial x^2} \right] = 0 . \tag{4.75}$$

According to Olavo [205], both the coupled equations (4.61) and (4.75), the so-called Madelung equations, lead to the Schrödinger equation (4.14) within the equivalence region between these equations.

We should note that (4.61) and (4.75) under the quantization conditions (4.72) and (4.74) are taken as the fundamental equations of quantum dynamics, while the Schrödinger equation is regarded as a mere mathematical consequence. The Olavo quantization method has been applied to obtain nonlinear Schrödinger equations [206] and elucidate the physical meaning of non-Boltzmann entropies [207] (e.g., Tsallis entropy [208]). Moreover, the relationships between his quantization method and the Feynman, Bohr–Sommerfeld and Nelson quantization schemes are studied critically in [203,205,207].

It is important to emphasize that, with regard to the quantization processes elaborated by Heisenberg, Dirac, Schrödinger, Feynman, and Nelson, the Olavo method possesses some quite attractive features (e.g., the concept of entropy) from a physical standpoint, not to mention a clear simplicity in the mathematical manipulations. Nevertheless, Olavo quantization is restricted to isolated systems, since the concept of entropy is not well-defined for dissipative systems in nonequilibrium thermodynamics [126,127]. However, it would be interesting to investigate the possibility of quantizing a Brownian particle in thermodynamic equilibrium, as described by the Markovian Fokker–Planck equation (3.128), using the Boltzmann entropy.

A highly relevant fact arising from the Olavo and Nelson quantization processes is that we do not need any Hamiltonian and/or Lagrangian formalism to quantize a determined classical physical system. In the next section we aim to continue in that same direction.

4.9 Dynamical Quantization

Recently, we have implemented a general method of quantization without using Hamiltonian and/or Lagrangian functions [40–44,209]. Since our starting point is the proper equations of motion, we have called this dynamical quantization. Our main objective below is to use this method to quantize the generalized Langevin equations (4.1) and (4.2) in two cases: first for a Markovian system (Sect. 4.9.1) and then for a non-Markovian system (Sect. 4.9.2) described by Fokker–Planck equations for two variables. In Sect. 4.9.3, we quantize anomalous Brownian motion.

4.9.1 Quantization of the Markovian Fokker–Planck Equation

Let us consider a Brownian particle described by the Markovian Langevin equations

$$\frac{\mathrm{d}P}{\mathrm{d}t} = -2\gamma P - \frac{\partial V}{\partial X} + b_1 \Phi(t) \,, \tag{4.76}$$

$$\frac{\mathrm{d}X}{\mathrm{d}t} = \frac{P}{m} \,, \tag{4.77}$$

with

$$\langle \Phi(t) \rangle = 0 \,, \tag{4.78}$$

$$\langle \Phi(t_1)\Phi(t_2) \rangle = 2\mathcal{D}^{1/3}\delta(t_1 - t_2) \,. \tag{4.79}$$

The diffusion coefficient $\mathcal{D}$ could denote the quantum nature of the thermal reservoir through

$$\mathcal{D} = m\gamma\omega\hbar \coth \frac{\omega\hbar}{2k_\mathrm{B}T} \,, \tag{4.80}$$

or, in the case $T \to 0$, through

$$\mathcal{D} = m\gamma\omega\hbar \,.$$

From (4.80), we encounter the expression in the classical domain $\hbar \to 0$,

$$\mathcal{D} = 2m\gamma k_\mathrm{B}T \,. \tag{4.81}$$

The Fokker–Planck equation for the probability density function $D_{XP} = D_{XP}(x, p, t)$ is given by

$$\frac{\partial}{\partial t} D_{XP} + \frac{p}{m}\frac{\partial}{\partial x} D_{XP} - \frac{\partial V(x,t)}{\partial x}\frac{\partial}{\partial p} D_{XP} = 2\gamma D_{XP} + 2\gamma p \frac{\partial}{\partial p} D_{XP} + \mathcal{D}\frac{\partial^2}{\partial p^2} D_{XP} \,. \tag{4.82}$$

Carrying out the Fourier transform

$$\chi(x, \eta, t) = \int_{-\infty}^{\infty} D_{XP}(x, p, t) \mathrm{e}^{\mathrm{i}p\eta} \mathrm{d}p \,, \tag{4.83}$$

the Fokker–Planck equation (4.82) becomes

$$\frac{\partial \chi}{\partial t} - \frac{\mathrm{i}}{m}\frac{\partial^2 \chi}{\partial \eta \partial x} + \mathrm{i}\eta \frac{\partial V}{\partial x}\chi + 2\gamma\eta \frac{\partial \chi}{\partial \eta} + \mathcal{D}\eta^2\chi = 0 \,. \tag{4.84}$$

In [40–42,44], we called the Fourier transform (4.83) the classical Wigner function. It is responsible for a novel representation of classical mechanics in which we derive an operator structure similar to quantum mechanics (see also Appendix E). Note that in the limit $\eta \to 0$, the classical Wigner function (4.83) turns out to be the marginal probability density $D_X(x, t)$. The exponential factor $\mathrm{e}^{\mathrm{i}p\eta}$ is a dimensionless term, and the variable η has the dimensions of a reciprocal linear momentum.

The change of variables from (x,η) to (x_1,x_2) given by

$$x=\frac{x_1+x_2}{2}\,, \qquad \eta=\frac{x_1-x_2}{\ell} \tag{4.85}$$

transforms (4.84) to

$$\mathrm{i}\ell\frac{\partial\chi}{\partial t}+\frac{\ell^2}{2m}\left(\frac{\partial^2\chi}{\partial x_1^2}-\frac{\partial^2\chi}{\partial x_2^2}\right)-\Big[V(x_1,t)-V(x_2,t)+O(x_1,x_2,t)\Big]\chi+\mathcal{I}\chi=0\,, \tag{4.86}$$

where

$$O(x_1,x_2,t)=-\sum_{n=3,5,7,\ldots}^{\infty}\frac{2^{1-n}}{n!}(x_1-x_2)^n\left(\frac{\partial}{\partial x_1}+\frac{\partial}{\partial x_2}\right)^n V(x_1,x_2,t) \tag{4.87}$$

and

$$\mathcal{I}=\gamma\mathrm{i}\ell(x_1-x_2)\left(\frac{\partial}{\partial x_1}-\frac{\partial}{\partial x_2}\right)+\frac{\mathrm{i}\mathcal{D}}{\ell}(x_1-x_2)^2\,. \tag{4.88}$$

From (4.85), we note that, if $\eta\ell$ is to have the dimensions of length, the arbitrary parameter ℓ must have the dimensions of action or angular momentum. Furthermore, because there exists an inverse Fourier transform for (4.83), viz.,

$$D_{XP}(x,p,t)=\frac{1}{2\pi}\int_{-\infty}^{\infty}\chi(x,\eta,t)\mathrm{e}^{-\mathrm{i}p\eta}\mathrm{d}\eta\,, \tag{4.89}$$

the Fokker–Planck equation (4.82) and (4.86) are equivalent.

Just as happens with the Kolmogorov stochastic equation (3.1), equation (4.86) is physically intractable due to the infinite number of terms. We thus let $\xi=x_1-x_2$ be a space interval, with $x_1>x_2$, so that it has the same infinitesimality property as (3.3), i.e.,

$$(x_1-x_2)^3\to 0\,. \tag{4.90}$$

As already seen in Chap. 3, Einstein [36] implicitly used the condition (4.90) to derive the diffusion equation for the probability density function. Recently, we used such an infinitesimality condition to derive quantum mechanical equations of motion [40–44,209]. Hence, we call (4.90) the Olavo–Einstein infinitesimality condition.

The Olavo–Einstein infinitesimality condition (4.90) together with the quantum limit

$$\ell\to\hbar=\frac{h}{2\pi}\,, \tag{4.91}$$

are the only conditions we use to quantize a classical system. Here h is the Planck constant, whose value is fixed by nature. [The numerical value of the Planck constant depends on an arbitrary choice of measurement units. In the MKSA (meter–kilogram–second–ampere) system, $\hbar$ has value $1.054592 \pm 0.000006 \times 10^{-34}$ MKSA units, while in another unit system we may find $\hbar = 1$ [210].] Because of (4.90) and (4.91), the equation of motion (4.86) becomes

$$\mathrm{i}\hbar\frac{\partial\rho}{\partial t} + \frac{\hbar^2}{2m}\left(\frac{\partial^2\rho}{\partial x_1^2} - \frac{\partial^2\rho}{\partial x_2^2}\right) - \Big[V(x_1,t) - V(x_2,t)\Big]\rho \qquad (4.92)$$
$$+\gamma\mathrm{i}\hbar(x_1 - x_2)\left(\frac{\partial\rho}{\partial x_1} - \frac{\partial\rho}{\partial x_2}\right) + \frac{\mathrm{i}\mathcal{D}}{\hbar}(x_1 - x_2)^2\rho = 0\ ,$$

where $\rho = \rho(x_1, x_2, t)$ is the von Neumann function (the density matrix or density operator in the coordinate representation). Equation (4.92) with diffusion constant $\mathcal{D}$ given by (4.81) is the Caldeira–Leggett equation found using Dirac quantization and the Feynman path integrals. Below, we compare our derivation with that performed by Caldeira and Leggett [130]:

- We have derived (4.92) without making any restrictions on the initial conditions. By contrast, as a consequence of the influence functional path-integral method of Feynman and Vernon [173] in the Caldeira–Leggett theory, the derivation is only obtained by supposing that the particle and environment are initially uncorrelated and by assuming the factorization of the von Neumann function in terms of variables of the particle and the thermal reservoir. This factorization assumption is not realistic and leads to non-physical results for the mean position, for instance [136,211–216]. Even though the Feynman technique can be adapted for more general initial conditions [133,136,211–213,217,218], our derivation from (4.92) shows that the Dirac and Feynman quantizations fail to reflect the underlying physics. Furthermore, due to the difficulty in manipulating the influence functionals, it is an extremely complex and ambiguous task to derive a single equation of motion such as (4.92) for arbitrary initial conditions and nonlinear external forces [133,216,219–223]. In contrast, the most advanced mathematical tool we employ in our quantization method is an elementary Fourier transform given by (4.83). [Incidentally, after repeating the calculations involving the influence functionals, Weiss ([136], p. 104) has found a term missing in the original work by Caldeira and Leggett [130]. Nevertheless, we cannot say whether such an error affects the final form of the master equation they obtained.]

- As the diffusion constant may be of either a classical nature (4.81) or a quantum nature (4.80) in our quantization process, we do not need to distinguish regimes of high and low temperatures and make specific assumptions about the nature of the environment. (We need only the friction and diffusion constants.) In the approach of Caldeira and Leggett, however, the thermal reservoir is treated quantum mechanically and their master equation is only valid within a quasi-classical realm of quantum mechanics characterized by high temperatures and by the Ohmic nature of the environment [224,228]. These restrictions are due to the Hamiltonian model required by Caldeira and Leggett. In contrast, we have shown that such a model is not needed to quantize a Brownian particle.
- In constrast with our (4.92), the original Caldeira–Leggett equation [130] (see also Sect. 4.6) is expressed in terms of a renormalized external potential. As our derivation has revealed, this feature is a mere artifact of the Hamiltonian model assumed in [130].
- When we consider the diffusion constant $\mathcal{D}$ as given by (4.80) in (4.92), we obtain the Markovian master equation found by Caldeira, Cerdeira and Ramaswamy [228] using the laborious Feynman technique and assuming the limit of very weak damping. As we have seen, this hypothesis is not necessary for deriving (4.92).
- Because the Caldeira–Leggett equation (4.92) is derived from the Fokker–Planck equation (4.82), it thereby inherits its Markovian characteristics. This property leads to serious physically undesirable results since $\rho(x_1, x_2, t)|_{x_1=x_2=x}$ turns out to have negative values [133,225–227].

In order to overcome the difficulties associated with the Markovian nature of the Caldeira–Leggett equation, we now wish to quantize a non-Markovian stochastic system using our quantization method.

4.9.2 Quantization of the Non-Markovian Fokker–Planck Equation

Our Brownian particle is described by the generalized Langevin equations

$$\frac{\mathrm{d}P}{\mathrm{d}t} = -\gamma P - \frac{\partial}{\partial X} V(X, t) + b_1(t)\Phi(t) , \tag{4.93}$$

$$\frac{\mathrm{d}X}{\mathrm{d}t} = \frac{P}{m} + b_2(t)\Phi(t) , \tag{4.94}$$

with

$$\langle \Phi(t_1) \rangle = \frac{D}{t_c} \mathrm{e}^{-t_1/t_c} , \tag{4.95}$$

$$\langle \Phi(t_1)\Phi(t_2) \rangle = \frac{D}{t_c} \mathrm{e}^{-(t_1+t_2)/2t_c} \delta(t_1 - t_2) , \tag{4.96}$$

$$\langle \Phi(t_1) \ldots \Phi(t_{2k+1}) \rangle = 0 \quad (k = 1, 2, \ldots) , \tag{4.97}$$

$$\langle \Phi(t_1) \ldots \Phi(t_{2k}) \rangle = 0 \quad (k = 2, 3, \ldots) . \tag{4.98}$$

The coefficients b_1 and b_2 are given by

$$b_1(t) = b_2(t) = D \left(1 - \mathrm{e}^{-t/t_c}\right) . \tag{4.99}$$

The non-Markovian Fokker–Planck equation generated by (4.93–4.99) has the form

$$\begin{aligned} \frac{\partial D_{XP}}{\partial t} &= -\frac{p}{m}\frac{\partial D_{XP}}{\partial x} - A(t)\frac{\partial D_{XP}}{\partial x} + \frac{\partial V}{\partial x}\frac{\partial D_{XP}}{\partial p} \\ &\quad + \gamma D_{XP} + \gamma p \frac{\partial D_{XP}}{\partial p} + B(t)\frac{\partial^2 D_{XP}}{\partial x^2} + C(t)\frac{\partial^2 D_{XP}}{\partial x \partial p} , \end{aligned} \tag{4.100}$$

where

$$A(t) = \frac{D^2}{m}\left(\mathrm{e}^{-t/t_c} - \frac{1}{2}\mathrm{e}^{-2t/t_c}\right) , \tag{4.101}$$

$$\begin{aligned} B(t) &= \frac{D^3}{m}\left(\mathrm{e}^{-t/t_c} - \mathrm{e}^{-2t/t_c} + \frac{1}{3}\mathrm{e}^{-3t/t_c}\right) \\ &\quad + \frac{D^3 t_c}{2m^2}\left(-\mathrm{e}^{-t/t_c} + \frac{1}{2}\mathrm{e}^{-2t/t_c} - \frac{1}{9}\mathrm{e}^{-3t/t_c}\right) , \end{aligned} \tag{4.102}$$

$$C(t) = \frac{2D^3}{m}\left(\mathrm{e}^{-t/t_c} - \mathrm{e}^{-2t/t_c} + \frac{1}{3}\mathrm{e}^{-3t/t_c}\right) . \tag{4.103}$$

The Fourier transform (4.83) and the change of variables (4.85) transform (4.100) to

$$\begin{aligned} 0 &= \mathrm{i}\ell\frac{\partial \chi}{\partial t} + \frac{\ell^2}{2m}\left(\frac{\partial^2 \chi}{\partial x_1^2} - \frac{\partial^2 \chi}{\partial x_2^2}\right) - \Big[V(x_1,t) - V(x_2,t)\Big]\chi \\ &\quad + O(x_1,x_2,t)\chi + \Big[\mathrm{i}\ell A(t) - (x_1 - x_2)C(t)\Big]\left(\frac{\partial \chi}{\partial x_1} + \frac{\partial \chi}{\partial x_2}\right) \\ &\quad + \frac{\mathrm{i}\ell}{2}\gamma(x_1 - x_2)\left(\frac{\partial \chi}{\partial x_1} - \frac{\partial \chi}{\partial x_2}\right) - \mathrm{i}\ell B(t)\left(\frac{\partial}{\partial x_1} + \frac{\partial}{\partial x_2}\right)^2 \chi . \end{aligned} \tag{4.104}$$

Using the quantization conditions (4.90) and (4.91), we obtain the non-Markovian master equation

$$
\begin{aligned}
0 = {} & \mathrm{i}\hbar\frac{\partial\rho}{\partial t} + \frac{\hbar^2}{2m}\left(\frac{\partial^2\rho}{\partial x_1^2} - \frac{\partial^2\rho}{\partial x_2^2}\right) - \Big[V(x_1,t) - V(x_2,t)\Big]\rho \\
& + \Big[\mathrm{i}\hbar A(t) - (x_1 - x_2)C(t)\Big]\left(\frac{\partial\rho}{\partial x_1} + \frac{\partial\rho}{\partial x_2}\right) \\
& + \frac{\mathrm{i}\hbar}{2}\gamma(x_1 - x_2)\left(\frac{\partial\rho}{\partial x_1} - \frac{\partial\rho}{\partial x_2}\right) - \mathrm{i}\hbar B(t)\left(\frac{\partial}{\partial x_1} + \frac{\partial}{\partial x_2}\right)^2\rho\,.
\end{aligned}
\tag{4.105}
$$

In the Markovian limit $t_\mathrm{c} \to 0$, equation (4.105) reduces to the dissipative von Neumann equation

$$
\begin{aligned}
0 = {} & \mathrm{i}\hbar\frac{\partial\rho}{\partial t} + \frac{\hbar^2}{2m}\left(\frac{\partial^2\rho}{\partial x_1^2} - \frac{\partial^2\rho}{\partial x_2^2}\right) \\
& - \Big[V(x_1,t) - V(x_2,t)\Big]\rho + \frac{\mathrm{i}\hbar}{2}\gamma(x_1 - x_2)\left(\frac{\partial\rho}{\partial x_1} + \frac{\partial\rho}{\partial x_2}\right)\,.
\end{aligned}
\tag{4.106}
$$

When the frictional constant vanishes, i.e., $\gamma \to 0$, we obtain the conservative von Neumann equation

$$
0 = \mathrm{i}\hbar\frac{\partial\rho}{\partial t} + \frac{\hbar^2}{2m}\left(\frac{\partial^2\rho}{\partial x_1^2} - \frac{\partial^2\rho}{\partial x_2^2}\right) - \Big[V(x_1,t) - V(x_2,t)\Big]\rho\,, \tag{4.107}
$$

which in turn reduces to the Schrödinger equation at point x_1, viz.,

$$
0 = \mathrm{i}\hbar\frac{\partial}{\partial t}\psi(x_1,t) + \frac{\hbar^2}{2m}\frac{\partial^2}{\partial x_1^2}\psi(x_1,t) - V(x_1,t)\psi(x_1,t)\,, \tag{4.108}
$$

and its complex conjugate at point x_2,

$$
0 = \mathrm{i}\hbar\frac{\partial}{\partial t}\psi^\dagger(x_2,t) - \frac{\hbar^2}{2m}\frac{\partial^2}{\partial x_2^2}\psi^\dagger(x_2,t) + V(x_2,t)\psi^\dagger(x_2,t)\,, \tag{4.109}
$$

since the von Neumann function ρ is factorized as

$$
\rho(x_1,x_2,t) = \psi(x_1,t)\psi^\dagger(x_2,t)\,. \tag{4.110}
$$

From the Schrödinger equation (4.108) at a generic point x, and expressing the Schrödinger function $\psi(x,t)$ as

$$
\psi(x,t) = R(x,t)\mathrm{e}^{\mathrm{i}S(x,t)/\hbar}\,, \tag{4.111}
$$

where R and S are in general functions dependent on $\hbar$, we obtain the Madelung representation of quantum mechanics [229] specified by the coupled equations

$$0 = \frac{\partial S}{\partial t} + \frac{1}{2m}\left(\frac{\partial S}{\partial x}\right)^2 - \frac{\hbar^2}{2mR}\frac{\partial^2 R}{\partial x^2} + V \,, \tag{4.112}$$

$$0 = \frac{\partial R^2}{\partial t} + \frac{1}{m}\frac{\partial}{\partial x}\left(R^2\frac{\partial S}{\partial x}\right) . \tag{4.113}$$

4.9.3 Quantization of Anomalous Brownian Motion

Rather than quantizing the Fokker–Planck equation in phase space, we can also quantize a Brownian particle described by the Smoluchowski equation in configuration space (see Appendix B). By way of example, in this section we restrict ourselves to the diffusion equation for anomalous or non-Einsteinian Brownian motion [43].

We start with the Kolmogorov stochastic equation (3.1). We introduce the Olavo–Einstein infinitesimality condition (4.90) and make use of the Pawula theorem. Next, we obtain the diffusion equation (3.9) at point x_1,

$$\frac{\partial f_1}{\partial t} = C\beta t^{\beta-1}\frac{\partial^2 f_1}{\partial x_1^2} \,, \tag{4.114}$$

and at point x_2,

$$\frac{\partial f_2}{\partial t} = C\beta t^{\beta-1}\frac{\partial^2 f_2}{\partial x_2^2} \,, \tag{4.115}$$

where $f_i = f(x_i, t)$ with $i = 1, 2$. We have chosen the anomalous diffusion coefficient given by (3.27). Multiplying (4.114) by f_2 and (4.115) by f_1 and adding the resulting equations, we get

$$\frac{\partial (f_1 f_2)}{\partial t} = C\beta t^{\beta-1}\left(\frac{\partial^2}{\partial x_1^2} + \frac{\partial^2}{\partial x_2^2}\right) f_1 f_2 \,. \tag{4.116}$$

We change the variables according to

$$x = \frac{x_1 + x_2}{\sqrt{2}} \,, \qquad \zeta = (x_1 - x_2)\sqrt{2} \,, \tag{4.117}$$

and arrive at the equation of motion for the function $\xi = \xi(x, \zeta, t) = f(x_1, t) f(x_2, t)$, viz.,

$$\frac{\partial \xi}{\partial t} = C\beta t^{\beta-1}\frac{\partial^2 \xi}{\partial x^2} + 4C\beta t^{\beta-1}\frac{\partial^2 \xi}{\partial \zeta^2} \,. \tag{4.118}$$

In the case $x_1 = x_2$, the function $f = \sqrt{\xi}$ is interpreted as a probability density function.

To quantize the diffusion equation (4.114), through (4.118), we perform the Fourier transform from (x, ζ)-space to (x, p)-space:

$$F(x,p,t) = \frac{1}{2\pi\hbar}\int_{-\infty}^{\infty} \xi(x,\zeta,t)\mathrm{e}^{\mathrm{i}(p-p_0)\zeta/\hbar}\mathrm{d}\zeta \,, \tag{4.119}$$

where p is the physical momentum $m\dot{x}$. Then (4.118) can be written

$$\frac{\partial F}{\partial t} = C\beta t^{\beta-1}\frac{\partial^2 F}{\partial x^2} - \frac{4D}{\hbar^2}\beta t^{\beta-1}(p-p_0)^2 F \,, \tag{4.120}$$

describing the motion of an anomalous Brownian particle in quantum phase space $(p, x; \hbar)$. We interpret $F = F(x,p,t)$ as a probability density function in quantum phase space $(x, p; \hbar)$, with the properties:

$$\int_{-\infty}^{\infty}\int_{-\infty}^{\infty} F\mathrm{d}x\mathrm{d}p = 1 \,, \quad \int_{-\infty}^{\infty} F\mathrm{d}x = g(p,t) \,, \quad \int_{-\infty}^{\infty} F\mathrm{d}p = h(x,t) \,.$$

The mean value of any physical quantity $A = A(p,x,t)$ is calculated by means of the expression

$$\int_{-\infty}^{\infty}\int_{-\infty}^{\infty} AF\mathrm{d}x\mathrm{d}p = \langle A\rangle \,.$$

To find the solution of (4.120), we choose the initial condition

$$F(x,p,t=0) = \frac{\sqrt{ab}}{\pi\hbar}\exp\left[-\frac{a}{\hbar}(p-p_0)^2 - \frac{b}{\hbar}(x-x_0)^2\right] \qquad (a,b>0) \,, \tag{4.121}$$

compatible with the Heisenberg relation $\sigma_x\sigma_p \geq \hbar/2$, where $\sigma_y = \sqrt{\langle y^2\rangle - \langle y\rangle^2}$. We then note that the parameters a and b associated with the unsharpness of the Gaussian (4.121) cannot vanish. [Equation (4.121) reduces to the classical case $\delta(x-x_0)\delta(p-p_0)$ as $\hbar \to 0$.] We thus find the normalized solution

$$F(x,p,t) = \frac{1}{2\pi}\sqrt{\frac{\mathcal{A}}{\mathcal{B}}}\exp\left[-\mathcal{A}(p-p_0)^2 - \frac{1}{4\mathcal{B}}(x-x_0)^2\right] \,, \tag{4.122}$$

where

$$\mathcal{A} = \frac{a}{\hbar} + \frac{4Ct^\beta}{\hbar^2} \,, \tag{4.123}$$

$$\mathcal{B} = \frac{\hbar}{4b} + Ct^\beta \,. \tag{4.124}$$

The solution (4.122) yields the results

$$\langle (x - x_0)^2 \rangle = \frac{\hbar}{2b} + 2Ct^\beta \,, \tag{4.125}$$

$$\langle (p - p_0)^2 \rangle = \frac{\hbar^2}{2(a\hbar + 4Ct^\beta)} \,, \tag{4.126}$$

$$\langle (p - p_0)^2 (x - x_0)^2 \rangle = \frac{\hbar^2}{4} \frac{\hbar + 4bCt^\beta}{ab\hbar + 4bCt^\beta} \,, \tag{4.127}$$

which reduce to the respective classical expressions in the limit $\hbar \to 0$. The quantum nature of the anomalous Brownian motion is revealed by the presence of Planck's constant in (4.125–4.127). Equation (4.127) shows that the uncertainty relation does not remain constant in time and depends on the anomalous parameter β. From (4.125–4.127), for $\beta = 1$, we recover ordinary Brownian motion with quantum effects.

We sum up the main results recently obtained from dynamical quantization without the need for Lagrangians and Hamiltonians:

- We have derived the following quantum mechanical equations of motion well-known in the literature:
  - the Caldeira–Leggett master equation [40,42],
  - the Caldeira–Cerdeira–Ramaswamy master equation [42],
  - the Schrödinger–Langevin equation [40,42],
  - the Caldirola–Kanai equation [40,42],
  - the Schrödinger equation,
  - the Madelung equations.
- We have also derived an operator structure for classical mechanics [42] (see also Appendix E).
- We have quantized nonlinear dissipative systems (without fluctuation), such as the van der Pol and Duffing systems [41].
- Quantization in generalized coordinates [205].
- Quantization of the non-Markovian Fokker–Planck equation (see Sect. 4.9.2).
- Quantization of anomalous Brownian motion (Sect. 4.9.3 and [43]).
- Quantization of the Smoluchowski equation (Appendix B).
- For isolated systems, we have compared our quantization process with the Dirac and Feynman procedures [44] (see also Appendixes C and D).

Mathematically, our method of quantization has shown that the main object of quantum theory is the von Neumann function $\rho(x_1, x_2, t)$. Even for isolated systems, the Schrödinger function plays a secondary role from a logical point of view, since it is derived from the von Neumann function via a procedure of variable separation $\rho(x_1, x_2, t) = \psi(x_1, t)\psi^\dagger(x_2, t)$.

Conceptually, we wish to emphasize that the trajectory concept of a particle in the classical domain is not destroyed by the Fourier transform (4.89) and conditions (4.90) and (4.91). That is, according to our quantization scheme, a quantum particle actually describes a trajectory, just like a planet or a Brownian particle.

Physically, we imagine the quantization process as a phenomenon happening during the transition from the classical domain to the quantum domain. It is analogous to a phase transition phenomenon. We notice that the Olavo–Einstein infinitesimality condition (4.90) is a precondition for both classical and quantum physics. In other words, it establishes a common range of applicability for both theories. We believe that a deeper theoretical and practical survey of quantization processes could reveal the true physical reason behind such an infinitesimality condition.

4.10 Quantum Physics Without Quantization Processes

By means of our quantization method we have shown that quantum physics arises from classical theory as a domain transition, a physical process analogous to a phase transition. Nevertheless, the more usual procedure is to avoid any quantization process and postulate ab initio the universality of the Schrödinger equation (for isolated systems) or the von Neumann equation for the density matrix or operator (for the case of open systems).

4.10.1 Isolated Systems

Schrödinger

Ab initio we should postulate [13,162] that the state of a quantum isolated system is specified by the function $\psi(x, t)$ whose time evolution is governed by the Schrödinger equation

$$\mathrm{i}\hbar\frac{\partial}{\partial t}\psi(x,t) + \frac{\hbar^2}{2m}\frac{\partial^2}{\partial x^2}\psi(x,t) - V(x,t)\psi(x,t) = 0 \,. \tag{4.128}$$

The complex-valued function $\psi(x, t)$ is interpreted as a probability amplitude, i.e., as a function giving rise to the concept of probability density:

$$\psi(x,t) \quad \Longrightarrow \quad |\psi(x,t)|^2 \,. \tag{4.129}$$

What is the physical meaning of the Schrödinger function ψ? Does the trajectory concept remain valid in the quantum world? Does an electron actually follow a trajectory around an atomic nucleus like a planet or a Brownian particle? How does the transition from ψ to $|\psi|^2$ occur? Over the last 80 years, several eminent physicists have made a great effort to answer these questions [13]. However, the debate seems to go on ad infinitum.

An easy way of sweeping such questions under the carpet is to adopt an operational point of view [230]: for a given physical system we solve the Schrödinger equation from certain initial and boundary conditions. Next we calculate $|\psi|^2$ and make some statistical predictions about the outcomes of measurements of physical quantities (e.g., position, energy). The trajectory concept, for example, turns out to be secondary in quantum theory. Only after a measurement process does such a concept begin to have physical significance, either in the wave form or in the particle form. What is important is that the predictions of the theory are in accord with the experimental facts. The rest is merely a question of interpretation.

Bohm

In order to explain the quantum phenomena Bohm, de Broglie et al. [15,230] formulated a quantum theory for isolated systems taking into account the Schrödinger equation (4.128) supplemented by the Madelung equations (4.112) and (4.113). In the Bohm approach, these three equations form the whole structure of quantum mechanics:

- A quantum particle possesses a wave ψ which guides it. The wave character of matter is described by the Schrödinger equation (4.110).
- The particle nature of matter is characterized by the quantum Hamilton–Jacobi equation(4.113), with the equation of motion

$$\dot{x} = \frac{1}{m}\frac{\partial S}{\partial x} \tag{4.130}$$

and the quantum potential

$$V_{\mathrm{Q}}(x,t) = -\frac{\hbar^2}{2mR}\frac{\partial^2 R}{\partial x^2} \,.$$

Both equations are determined by the wavefunction (4.111).

- The probabilistic character of the Bohm theory is specified by the continuity equation (4.113), since

$$R^2 = |\psi|^2 . \qquad (4.131)$$

Here the probability concept plays a secondary role in the theory. It is derived from the Schrödinger function.

Within the hypothesis of the universality of quantum mechanics for isolated systems, the book by Grib and Rodrigues Jr. [13] provides an excellent study exhibiting the advantages and shortcomings underlying the various interpretations (Copenhagen interpretation, many worlds, the Broglie–Bohm approach, quantum logic, decoherence due to environmental influence, and the histories approach).

4.10.2 Open Systems

The time evolution of an isolated quantum system may alternatively be represented by the von Neumann equation

$$\mathrm{i}\hbar \frac{\mathrm{d}}{\mathrm{d}t}\hat{\rho} = [\hat{H}, \hat{\rho}] , \qquad (4.132)$$

where $\hat{\rho}$ is the density operator and $\hat{H}$ the Hamiltonian operator. In technical terms, one says that the dynamics of an isolated system is generated by the one-parameter group of unitary transformations in Hilbert space. In this representation of quantum mechanics, (4.132) can be reduced to the Schrödinger equation by considering

$$\langle x_1|\hat{\rho}|x_2\rangle = \langle x_1|\psi\rangle\langle\psi|x_2\rangle = \psi(x_1,t)\psi^\dagger(x_2,t) .$$

Let us consider a system S coupled to an environment R, such that the total system $T = S + R$ can be treated as isolated, i.e.,

$$\mathrm{i}\hbar \frac{\mathrm{d}}{\mathrm{d}t}\hat{\rho}_{(S+R)} = [\hat{H}_{(S+R)}, \hat{\rho}_{(S+R)}] . \qquad (4.133)$$

Eliminating the environment variables under the hypothesis of weak coupling and Markovian evolution, Lindblad [231] obtained the master equation describing the non-Hamiltonian or open system S:

$$\mathrm{i}\hbar \frac{\mathrm{d}}{\mathrm{d}t}\hat{\rho} = [\hat{H}, \hat{\rho}] + \frac{\mathrm{i}}{2}\sum_k \left([\hat{L}_k\hat{\rho}, \hat{L}_k^\dagger] + [\hat{L}_k, \hat{\rho}\hat{L}_k^\dagger] \right) , \qquad (4.134)$$

where $\hat{L}_k$ are the Lindblad operators which contain all the effects of the environment R on the system S. The form of the Lindblad equation

(4.134) has the property of preserving the positivity of $\hat{\rho}$ for any initial condition. The equation of motion (4.134) is generated by a semigroup of transformations introducing irreversibility. Notice that (4.134) is not reducible to the deterministic Schrödinger equation, since $\hat{\rho} \neq |\psi\rangle\langle\psi|$.

Even though it is mathematically consistent, the Lindblad theory does not yield a method of construction for the operators $\hat{L}_k$. Specifically, it is straightforward to arrive at Markovian master equations of the form (4.134) leading to physically unacceptable results [232]. Therefore, it is important to set up phenomenological criteria for the validity of Markovian master equations. Without these criteria, the Lindblad equations are devoid of any physical significance. (Here we wish to point out that our quantization method presented in Sect. 4.9 does possess all the phenomenological content needed to guarantee a physical meaning for our in general non-Markovian quantum master equations.)

Summarizing, an axiomatic approach such as the Lindblad approach for treating the universality of quantum mechanics in terms of the von Neumann equation is not sufficient to apprehend the true physics of open systems. Recent controversies in this field [136,226,233–239] seem to support our point of view.

5 Classical Limit of Quantum Physics

5.1 Motivation: The Problem of the Classical Limit

For open systems, the transition from classical physics to quantum physics is carried out by means of a quantization process taking into account the effects caused by the appearance of the Planck constant. In turn the classical limit is a physical phenomenon characterized by a dequantization process in which the Planck constant also plays a fundamental role. Mathematically, we can represent such a physical situation through the limiting procedure

$$\hbar \to 0\,, \tag{5.1}$$

although physically $\hbar$ is a constant fixed by nature. We make the following requirements to ensure that the mathematical limit (5.1) makes some physical sense:

- The limit $\hbar \to 0$ should be a mathematically well-defined procedure. For example, taking $\hbar \to 0$ directly for the differential momentum operator
  $$\hat{p} = \frac{\hbar}{\mathrm{i}}\frac{\partial}{\partial x}$$
  leads to the nonsensical identity $\hat{p} = 0$, since the number zero is not an operator. As another example, the classical limit of the fine-structure constant
  $$\alpha = \frac{e^2}{\hbar c}\,,$$
  where $-e$ is the charge of the electron and c the velocity of light, gives rise to a divergence as $\hbar \to 0$. To avoid these difficulties, we must elaborate a method to make sure that the notion of classical limit makes sense.
- How should we describe the underlying physics during the domain transition quantum $\to$ classical?

Elucidating the second point remains a delicate problem as it involves investigating the quantization–dequantization process. Hence in the present chapter we restrict ourselves to the first item above. To this end, in Sects. 4.2–4.7, we present the following methods discussed in the literature: the Bohr correspondence principle (Sect. 4.2), the Ehrenfest theorem (Sect. 4.3), the WKB method (Sect. 4.4), the Feynman method (Sect. 4.5), the decoherence approach (Sect. 4.6), and the Bohm quantum potential (Sect. 4.7). Section 4.8 is devoted to our classical limiting method.

5.2 Bohr Correspondence Principle

According to the Bohr correspondence principle, as presented in textbooks, quantum theory reduces to classical theory in the limit of large quantum numbers

$$n \to \infty \, . \tag{5.2}$$

This limit has worked well historically, in the Bohr atomic model. Here the frequency of light emitted from the state n to $n - n'$ is given by the expression [241]

$$\omega_{n,n-n'} = \frac{me^4}{2\hbar^3}\left(\frac{1}{n^2} + \frac{2n'}{n^3} + \cdots - \frac{1}{n^2}\right) \, .$$

In the limit $n \to \infty$, we obtain the orbital frequency

$$\omega_{n,n-n'} = n' \frac{me^4}{n^3\hbar^3} \, .$$

However, other physical systems such as the harmonic oscillator [242] violate the Bohr correspondence principle. Recent discussions about the inadequacy of Bohr's correspondence principle (5.2) to connect quantum and classical physics can be found in [243–245].

The important point to be stressed is that the Bohr procedure does not make any allusion to the classical limit of quantum dynamics.

5.3 Ehrenfest Theorem

From the Schrödinger equation one derives the Ehrenfest equations [246]

$$\frac{\mathrm{d}\langle p\rangle_\psi}{\mathrm{d}t} = -\left\langle \frac{\partial V(x,t)}{\partial x} \right\rangle_\psi \, , \tag{5.3}$$

$$\frac{\mathrm{d}\langle x\rangle_\psi}{\mathrm{d}t} = \frac{1}{m}\langle p\rangle_\psi \,. \tag{5.4}$$

The average value $\langle A\rangle$ of a physical quantity $A(x,p,t)$ is calculated using the equation

$$\langle A\rangle_\psi = \int_{-\infty}^{\infty} \psi^\dagger \hat{A}\psi \mathrm{d}x \tag{5.5}$$

for the operator $\hat{A} = A(\hat{x},\hat{p},t)$, where ψ is normalized so that $\int_{-\infty}^{\infty}\psi^\dagger\psi\mathrm{d}x = 1$. Expanding the potential function $V = V(x,t)$ around the mean $\langle x\rangle_\psi$, differentiating with respect to x, and evaluating the mean $\langle \partial V/\partial x\rangle_\psi$, we arrive at

$$\left\langle\frac{\partial V(x,t)}{\partial x}\right\rangle_\psi = \frac{\partial}{\partial x}V(\langle x\rangle_\psi,t) + \sum_{k=2}^{\infty}\frac{1}{k!}\left\langle(x-\langle x\rangle_\psi)^k\right\rangle_\psi \frac{\partial^{k+1}}{\partial x^{k+1}}V(\langle x\rangle_\psi,t)\,. \tag{5.6}$$

Supposing that all the moments vanish, i.e.,

$$\left\langle(x-\langle x\rangle_{\psi_{\mathrm{EP}}})^k\right\rangle_{\psi_{\mathrm{EP}}} \to 0 \qquad (k=2,3,\ldots)\,, \tag{5.7}$$

for a given sufficiently localized function $\psi = \psi_{\mathrm{EP}}$ (Ehrenfest packet $|\psi_{\mathrm{EP}}|^2$), we obtain

$$\left\langle\frac{\partial V(x,t)}{\partial x}\right\rangle_{\psi_{\mathrm{EP}}} = \frac{\partial}{\partial x}V(\langle x\rangle_{\psi_{\mathrm{EP}}},t)\,. \tag{5.8}$$

Inserting (5.8) into (5.3), the Ehrenfest equations become

$$\frac{\mathrm{d}\langle p\rangle_{\psi_{\mathrm{EP}}}}{\mathrm{d}t} = -\frac{\partial}{\partial x}V(\langle x\rangle_{\psi_{\mathrm{EP}}},t)\,, \tag{5.9}$$

$$\frac{\mathrm{d}\langle x\rangle_{\psi_{\mathrm{EP}}}}{\mathrm{d}t} = \frac{1}{m}\langle p\rangle_{\psi_{\mathrm{EP}}}\,. \tag{5.10}$$

One then says that the Newtonian equations for the mean values $\langle x\rangle_{\psi_{\mathrm{EP}}}$ and $\langle p\rangle_{\psi_{\mathrm{EP}}}$ are obtained as a classical limit of the Schrödinger equation. Ehrenfest [246] considered two examples:

- For a particle of mass $m = 1$ g with an initial dispersion of 10^{-3} cm, the dispersion time or Ehrenfest time is 10^{21} s. This result is quite difficult to test experimentally. One therefore justifies the use of the trajectory concept in the classical domain, since ψ_{EP} is always well localized.
- For a particle of mass $m = 1.7\times10^{-24}$ g and initial dispersion 10^{-8} cm the Ehrenfest time is equal to 10^{-13} s. In the quantum domain, there is no place for the concept of trajectory.

Let us point out the applicability conditions of the Ehrenfest theorem:

- It depends on the mass of the system: the spreading of the Schrödinger function is negligible for large masses.
- It depends on the type of solution of the Schrödinger equation: the superposition of ψ violates the Ehrenfest theorem (5.9) and (5.10).

Recently, Ballentine et al. [247] have shown that the condition (5.7) for the validity of the Ehrenfest theorem is neither necessary nor sufficient for characterizing the classical limit of quantum mechanics. On the one hand, for potential functions of the form $V = a + bx + cx^2$, with a, b, c constants, the condition (5.7) is not sufficient to guarantee the classical nature of these systems. Let us consider a harmonic oscillator as an example. Even though the mean behavior given by (5.9) and (5.10) exactly simulates classical behavior for any ψ, it is well known that both the quantum and classical oscillators are not physically identical because of the zero point energy. On the other hand, there exist physical systems (e.g., a particle between reflecting walls and driven quartic oscillator) reaching the classical domain and at the same time violating the condition (5.7).

In summary, the condition (5.7) for applying Ehrenfest's theorem is very restrictive. Consequently, such a theorem is not reliable for investigating the problem of the classical limit of quantum dynamics in a general way.

5.4 WKB Method

According to the WKB method [248–250], we insert the special wave function

$$\psi(x,t) = \mathrm{e}^{(\mathrm{i}/\hbar)\mathcal{S}(x,t)} \tag{5.11}$$

into the Schrödinger equation to give

$$-\frac{\mathrm{i}\hbar}{2m}\frac{\partial^2 \mathcal{S}(x,t)}{\partial x^2} + \frac{1}{2m}\left[\frac{\partial \mathcal{S}(x,t)}{\partial x}\right]^2 + V(x,t) = -\frac{\partial \mathcal{S}(x,t)}{\partial t} . \tag{5.12}$$

We insert the so-called WKB expansion

$$\mathcal{S}(x,t) = S(x,t) + \frac{\hbar}{\mathrm{i}} S_1(x,t) + \left(\frac{\hbar}{\mathrm{i}}\right)^2 S_2(x,t) + \cdots \tag{5.13}$$

into (5.12), with $S(x,t)$ independent of $\hbar$. We then find the Hamilton–Jacobi equation for the function $S(x,t)$, viz.,

$$\frac{1}{2m}\left[\frac{\partial S(x,t)}{\partial x}\right]^2 + V(x,t) = -\frac{\partial S(x,t)}{\partial t} \,, \tag{5.14}$$

as the classical limit of the Schrödinger equation, provided that there are asymptotics

$$\lim_{\hbar\to 0} \hbar^n S_n \sim 0 \qquad (n = 1, 2, \dots) \,. \tag{5.15}$$

The Schrödinger function (5.11) constructed with the solutions of the Hamilton–Jacobi equation (5.14) is said to be a semiclassical function. On the one hand, it is unsuitable for describing classical systems because the WKB function ψ_{WKB} defined by

$$\lim_{\hbar\to 0} \psi(x,t) \sim \psi_{\mathrm{WKB}} = \mathrm{e}^{(\mathrm{i}/\hbar)S(x,t)} \tag{5.16}$$

is not a solution of a classical equation of motion. On the other hand, (5.16) does not satisfy the Schrödinger equation. The special function (5.16) is valid only in the semiclassical realm of quantum mechanics.

Even though the WKB method correctly provides the classical limit of the Schrödinger equation, it suffers from some difficulties:

- The superposition of WKB functions, viz.,

  $$\psi' = \frac{1}{2}\mathrm{e}^{\mathrm{i}S/\hbar} + \frac{1}{2}\mathrm{e}^{-\mathrm{i}S/\hbar} = \cos\frac{S}{\hbar} \,, \tag{5.17}$$

  does not lead to the classical Hamilton–Jacobi equation (5.14). The WKB method is therefore incompatible with the superposition principle of quantum mechanics (isolated systems).
- Another difficulty is that there are physical systems whose solutions cannot be put in the form (5.11) with the expansion (5.13).

The restrictions underlying the validity conditions of the WKB method indicate its lack of generality for calculating the classical limit of quantum dynamics.

5.5 Feynman Method

In Feynman's representation of quantum mechanics, one replaces the Schrödinger equation by the integral equation [11]

$$\psi(x,t) = \int_{-\infty}^{\infty} K(x,t;x_0,t_0)\psi(x_0,t_0)\mathrm{d}x_0 \,, \tag{5.18}$$

where Feynman's kernel $K(x,t;x_0,t_0)$ is given by

$$K(x,t;x_0,t_0) = N \int_{x_0,t_0}^{x,t} \mathrm{e}^{(\mathrm{i}/\hbar)S[x(t)]} \mathrm{d}[x(t)] \,. \tag{5.19}$$

N denotes a normalization constant and the sum is over all possible quantum trajectories or paths connecting points (x_0,t_0) and (x,t). If we suppose that the quantum function $S[x(t)]$ may be expanded according to the WKB expansion (5.13),

$$S[x(t)] = S_{\mathrm{cl}}(x,t;x_0,t_0) + O(\hbar) \,, \tag{5.20}$$

then in the classical limit $\hbar \to 0$, we recover the classical action function

$$S_{\mathrm{cl}}(x,t;x_0,t_0) = \int_{x_0,t_0}^{x,t} L(x,\dot{x},t)\mathrm{d}t = \int_{x_0,t_0}^{x,t} \left(\frac{m\dot{x}^2}{2} - V\right) \mathrm{d}t \,, \tag{5.21}$$

from which we obtain the Lagrangian equations of motion from the principle of least action.

Under the assumption (5.20), one says that, in Feynman's integral representation of quantum theory, the classical limit of the Schrödinger equation is given by the Lagrange equations [11].

In the semiclassical realm of quantum mechanics, the propagator (5.20) built with the help of the classical action (5.22) turns out to be [181]

$$\lim_{\hbar \to 0} K(x,t;x_0,t_0) \sim K_{\mathrm{sc}} = \text{smooth function} \times \mathrm{e}^{(\mathrm{i}/\hbar)S_{\mathrm{cl}}} \,. \tag{5.22}$$

This semiclassical propagator (5.22) is not valid for an arbitrary potential. Moreover, it is unsuitable for calculating the Schrödinger function ψ in terms of simple quadratures in the long time regime [180].

5.6 Decoherence

An isolated system is described by the Schrödinger equation, whose main characteristic is the superposition principle. The latter may, for example, be expressed in terms of wave functions by

$$\psi(x,t) = \psi_1(x,t) + \psi_2(x,t) \,, \tag{5.23}$$

or, in terms of the density matrix,

$$\rho(x,t) = \rho_1(x,t) + \rho_2(x,t) + \rho_{\mathrm{int}}(x,t) \,, \tag{5.24}$$

where $\rho_1(x,t) = \psi_1^\dagger(x,t)\psi_1(x,t)$, $\rho_2(x,t) = \psi_2^\dagger(x,t)\psi_2(x,t)$. The interference term is given by $\rho_{\mathrm{int}}(x,t) = \psi_1^\dagger(x,t)\psi_2(x,t) + \psi_2^\dagger(x,t)\psi_1(x,t)$.

In the quantum domain, the coherence of the solutions (5.23), represented by the presence of the interference term in (5.24), predicts physically different states for the maser mode of an ammonia molecule, for instance. Nevertheless, in the classical domain, there exist physical systems that do not present any property based on the coherence of ψ functions (e.g., the chiral states of a sugar molecule or the moon being in different places). Could quantum theory also be applied to the classical domain?

Under the hypothesis of universality of quantum mechanics, the decoherence approach [157,251,252] aims to explain by means of a physical mechanism how the coherence of ψ turns out to be unobservable in the classical domain (H.D. Zeh [157]): "the theory of decoherence is to explain the difference in appearance between the quantum and the classical under the assumption of a universally valid quantum theory." A classical limiting process compatible with the superposition of quantum mechanics is based on a loss of coherence (or decoherence criterion):

$$\rho_{\mathrm{int}}(x,t) \to 0\,. \tag{5.25}$$

According to the decoherence approach, in the quantum domain, systems are isolated from their surroundings, whereas in the classical domain, one supposes that physical systems are open quantum systems, no longer described by the Schrödinger equation. The fundamental equation becomes the von Neumann master equation including dissipation and fluctuation effects which are responsible for eliminating or destroying the coeherence between quantum states (5.25) in the classical domain.

To illustrate the application of condition (5.25), let us consider a situation where the decoherence process is governed by the Caldeira–Leggett factor [13,253,254]

$$I(t) = \exp\left[-\frac{m\omega}{4\hbar}(x_1 - x_2)^2(1 - \mathrm{e}^{-\lambda t})\right]\,. \tag{5.26}$$

For a pendulum of mass $m = 1$ g, period $2\pi/\omega = 1$ s, damping time $1/\lambda = 60$ s, initial distance $x_1 - x_2 = 10^{-6}$ m and time 10^{-9} s, we find the value $I(t) = \exp(-10^5)$. This is too tiny to be observable in practice. Thus, assuming that the physical bodies in the classical domain are always coupled to an environment, one obtains the result that quantum mechanics provides all the properties of classical systems by means of statistical mixtures. Decoherence therefore explains the classicality of the physical world.

In order to evaluate the status of decoherence as a procedure for connecting quantum and classical physics, we wish to make the following remarks:

- Despite the universality of the Schrödinger equation, the process of openness

$$\text{isolated systems} \quad \Longrightarrow \quad \text{open systems}\ ,$$

  that is, the elimination of the degrees of freedom of the environment for deriving quantum master equations, is not described by the quantum theory. In fact, as we have shown in Sect. 4.9, the Caldeira–Leggett equation assumes the validity of the Langevin equations in the classical domain from which it is derived [255]. No Hamiltonian model, i.e., no scheme for integrating out the environment, has been required there. One way of trying to save the universal validity of quantum mechanics is to base the theory of decoherence on the Lindblad master equations. However, this axiomatic procedure does not overcome certain difficulties such as the need to introduce some phenomenological criteria in an ad hoc manner in order to ensure that the Markovian Lindblad formalism has physical significance (see Sect. 4.10). We think that this lack of universality of quantum mechanics introduces at least a small logical incoherence into the decoherence approach.
- How does the trajectory concept emerge from a theory that does not possess such a concept? Answering this question involves solving the problem of quantum measurement. From an objective perspective, the decoherence approach is insufficient to solve it [251,255,256], while from a subjective angle, its relationship with the von Neumann esotoric ego still stands as a mystery [13,251].
- How can we calculate the classical limit of quantum mechanical equations of motion such as the Caldeira–Leggett equation, the Schrödinger equation, and our non-Markovian master equation? Perhaps such questions are meaningless within the decoherence approach, since no quantum Brownian motion (open system) seems to exist in the quantum domain.

In short, the major difficulty in establishing a general theory of decoherence lies in the hypothesis concerning the universal validity of quantum mechanics. Thus, although eminent physicists share the same opinion that the decoherence approach "is nowadays generally accepted by the respectable community" [136], we should point out that the hypothesis of universality of the Schrödinger equation, according to which all the physical systems are assumed to be isolated, gives rise unavoidably to the following paradoxical situation:

$$\text{coherence} \quad \Longrightarrow \quad \text{decoherence} \quad \Longrightarrow \quad \text{incoherence}\,.$$

5.7 Bohm Quantum Potential

Using the Madelung representation of quantum mechanics [229], one obtains the quantum Hamilton–Jacobi equation (see Sect. 4.10)

$$\frac{\partial \mathcal{S}}{\partial t} + \frac{1}{2m}\left(\frac{\partial \mathcal{S}}{\partial x}\right)^2 + Q(x,t) + V(x,t) = 0\,, \tag{5.27}$$

where

$$Q(x,t) = -\frac{\hbar^2}{2m}\frac{1}{R}\frac{\partial^2 R}{\partial x^2} \tag{5.28}$$

is the Bohm quantum potential. The functions R and $\mathcal{S}$ are related to the wave function ψ by means of the relation

$$\psi(x,t) = R(x,t)\mathrm{e}^{(\mathrm{i}/\hbar)\mathcal{S}(x,t)}\,. \tag{5.29}$$

On the basis of the assumption of universality of quantum mechanics in Bohm's ontological interpretation, the particles actually follow a (quantum) trajectory independent of any measurement apparatus. Rather than considering $\hbar \to 0$, since in (5.28) R is in general a function depending on $\hbar$, the correct procedure for connecting the quantum domain to the classical one is by means of the limit [15,230,257]

$$Q(x,t) \to 0\,, \tag{5.30}$$

so that (5.27) reduces to the classical Hamilton–Jacobi equation. When it exists, the condition (5.30) establishes a classical limiting process dependent on the choice of quantum states ψ.

When the quantum potential is identically zero, i.e., $Q = 0$, the quantum trajectories coincide with the classical ones, just as happens in the case of Ehrenfest's theorem for potential $V = a + bx + cx^2$. In contrast, in the Madelung representation, there are a great many potentials (a determined class of central potentials) for which some quantum states follow a trajectory coinciding with the classical one [258,259]. It should be noted, however, that this simulation does not characterize the classical limit of quantum dynamics.

By way of example, we consider three cases illustrating the use of the Bohm condition $Q \to 0$.

Semiclassical Quantum Mechanics

In the semiclassical region of quantum mechanics, we may use the WKB function for stationary states of energy $E > V$ in the form [230]

$$\psi_{\mathrm{WKB}} = \frac{A}{[2m(E-V)]^{1/4}} \exp\left(\frac{1}{\hbar}\left\{\int_{-\infty}^{x} [2m(E-V)]^{1/2}\mathrm{d}x - Et\right\}\right) , \tag{5.31}$$

where A is a constant. The quantum potential generated by (5.31) is given by

$$Q_{\mathrm{WKB}} = -\frac{\hbar^2}{2m}(E-V)^{1/4}\frac{\mathrm{d}^2}{\mathrm{d}x^2}(E-V)^{-1/4} . \tag{5.32}$$

The validity conditions of the WKB function (5.31) imply that the quantum potential has the following behavior:

$$|Q_{\mathrm{WKB}}| \ll (E-V) \tag{5.33}$$

and

$$\left|\frac{\mathrm{d}Q_{\mathrm{WKB}}}{\mathrm{d}x}\right| \ll \left|\frac{\mathrm{d}V}{\mathrm{d}x}\right| , \tag{5.34}$$

thus justifying the use of the condition (5.30).

Free Particle $V = 0$

Let us consider the Schrödinger equation for a free particle. As a solution we have

$$\psi' = A\mathrm{e}^{\mathrm{i}(px - p^2t/2m)/\hbar} \quad \text{or} \quad \psi'' = A\mathrm{e}^{-\mathrm{i}(px + p^2t/2m)/\hbar} , \tag{5.35}$$

where A and p are real constants. The quantum potential (5.28) associated with ψ' or ψ'' is zero and $p = \partial S/\partial x$, whereas the superposition $\psi = (\psi' + \psi'')/\sqrt{2}$ leads to $Q = p^2/2m$. Then according to the Bohm approach, the classical limit of the Schrödinger equation describing a free particle does not occur via the criterion $Q \to 0$ for superposed wave functions.

Harmonic Oscillator $V = m\omega^2 x^2/2$

The stationary solutions for this physical system given by

$$\psi_n(x) = c_n H_n(\xi x) \mathrm{e}^{-\xi^2 x^2/2} , \tag{5.36}$$

where $\xi = (m\omega/\hbar)^{1/2}$, $c_n = \xi/(\pi^{1/2} 2^n n!)^{1/2}$ and H_n are the Hermite polynomials, lead to the quantum potential

$$Q = \hbar\omega \left(n + \frac{1}{2} \right) - \frac{1}{2} m\omega^2 x^2 . \tag{5.37}$$

This differs from zero (even taking $\hbar \to 0$). Consequently, the quantum harmonic oscillator does not possess any classical limit when it is described by stationary states. Furthermore, the Bohm theory predicts a harmonic oscillator with $p = 0$ and total energy equal to $\hbar\omega(n + 1/2)$. Alternatively, if we impose ab initio the condition (5.30), we obtain the region where the classical limit exists from (5.37):

$$x = \pm \sqrt{\frac{2\hbar}{m\omega} \left(n + \frac{1}{2} \right)} .$$

However, this procedure seems to be incorrect. According to Holland [230], we first have to find quantum states for which the criterion (5.30) is satisfied. To this end we construct a coherent packet whose width is negligible compared with the amplitude of the oscillation. The condition $Q \to 0$ therefore depends on the wave function ψ for the system. Consequently, in Bohm's theory not all quantum mechanical systems have a classical limit.

On investigating the motion of a particle elastically reflected from a one-dimensional infinite potential barrier, Holland [260] shows that the quantum trajectories are not able to generate the full variety of the valid classical trajectories. Thus, starting with the hypothesis of universality of quantum theory (describing isolated systems), assumed by Bohm [15], Holland arrives at the following conclusion [260]: "there exist physically realisable classical systems that can never be reached as the limit of some quantum systems. Not all of classical mechanics where it is correct can be a special case of quantum mechanics." Even though that breakdown of the universality hypothesis is considered as "one of the most important new insights of the de Broglie–Bohm model" [260], it should be emphasized that "[classical systems that are not the limit of any quantum system] would not be good news for those who would like to consider Bohmian mechanics as the fundamental theory in both the classical and quantum domains" [261].

In short, according to our view the Bohm–de Broglie theory suffers from two defects:

- It is based on the hypothesis of the universal validity of quantum mechanics. In contrast, ontological characteristics can be carried to the quantum domain by means of our quantization process (Sect. 4.9).
- The vanishing of the quantum potential is not a necessary criterion for characterizing the classical limit of quantum mechanics, as we have already shown in [47].

5.8 Our Classical Limiting Process

According to our quantization method for open systems presented in Sect. 4.9, the von Neumann representation is the fundamental description of quantum mechanics. On the one hand, this means that we can arrive at the phase-space representation from it by means of a Fourier transform. On the other hand, provided the systems are isolated, we may obtain the Schrödinger representation from a single-variable separation $\rho(x_1, x_2, t) = \psi(x_1, t)\psi^\dagger(x_2, t)$:

$$\begin{array}{c}\text{phase-space representation}\\ \uparrow \\ \text{von Neumann representation}\\ \downarrow \\ \text{Schrödinger representation}\end{array}$$

In [41,42,44–47], we worked out a method for calculating the classical limit of the quantum dynamics of open systems without needing to take $\hbar \to 0$ directly in the solutions of the quantum mechanical equations of motion. Below we begin by presenting this method for isolated systems described by the Schrödinger equation (configuration space) or by the equations of motion in phase space (Wigner, Kirkwood, Husimi). We then calculate the classical limits of the Caldeira–Leggett equation and our non-Markovian master equation (4.105).

5.8.1 Isolated Systems

Classical Limit in Configuration Space

To make use of the limit $\hbar \to 0$, our main idea is to perform the unitary transformation

$$\psi'(x,t) = \mathrm{e}^{-(\mathrm{i}/\hbar)\xi(x,t)}\psi(x,t) , \tag{5.38}$$

so that the Schrödinger equation can be written as

$$\begin{aligned} 0 = &\frac{-\hbar^2}{2m}\frac{\partial^2\psi'}{\partial x^2} - \frac{\mathrm{i}\hbar}{2m}\left(\frac{\partial^2\xi}{\partial x^2}\psi' + 2\frac{\partial\xi}{\partial x}\frac{\partial\psi'}{\partial x} + 2m\frac{\partial\psi'}{\partial t}\right) \\ &+ \left[\frac{1}{2m}\left(\frac{\partial\xi}{\partial x}\right)^2 + V + \frac{\partial\xi}{\partial t}\right]\psi' . \end{aligned} \tag{5.39}$$

Using (5.38), we can obtain the classical limit of the operators $\hat{p} = -\mathrm{i}\hbar\partial/\partial x$, $\hat{p}^2 = -\hbar^2\partial^2/\partial x^2$, and $\hat{E} = \mathrm{i}\hbar\partial/\partial t$ as follows:

$$\lim_{\hbar\to 0}\hat{p}' = \lim_{\hbar\to 0}\left[\mathrm{e}^{-\mathrm{i}\xi/\hbar}\left(-\mathrm{i}\hbar\frac{\partial}{\partial x}\right)\mathrm{e}^{\mathrm{i}\xi/\hbar}\right] = \frac{\partial\xi}{\partial x} \equiv p , \tag{5.40}$$

$$\lim_{\hbar\to 0}\hat{p}'^2 = \lim_{\hbar\to 0}\left[\mathrm{e}^{-\mathrm{i}\xi/\hbar}\left(-\mathrm{i}\hbar\frac{\partial}{\partial x}\right)^2\mathrm{e}^{\mathrm{i}\xi/\hbar}\right] = \left(\frac{\partial\xi}{\partial x}\right)^2 \equiv p^2 , \tag{5.41}$$

$$\lim_{\hbar\to 0}\hat{E}' = \lim_{\hbar\to 0}\left[\mathrm{e}^{-\mathrm{i}\xi/\hbar}\left(\mathrm{i}\hbar\frac{\partial}{\partial t}\right)\mathrm{e}^{\mathrm{i}\xi/\hbar}\right] = -\frac{\partial\xi}{\partial t} \equiv E . \tag{5.42}$$

We now introduce the eigenvalue equation

$$\left(\frac{\hat{p}^2}{2m} - \hat{E}\right)\psi = \left(\frac{p^2}{2m} - E\right)\psi . \tag{5.43}$$

Using the transformation (5.38) and the relations (5.40–5.42), equation (5.43) becomes

$$-\frac{\hbar^2}{2m}\frac{\partial^2\psi'}{\partial x^2} - \frac{\mathrm{i}\hbar}{2m}\frac{\partial^2\xi}{\partial x^2}\psi' + \frac{\hbar}{\mathrm{i}m}\frac{\partial\xi}{\partial x}\frac{\partial\psi'}{\partial x} - \mathrm{i}\hbar\frac{\partial\psi'}{\partial t} = 0 . \tag{5.44}$$

Inserting (5.44) into (5.39), we arrive at

$$\left[\frac{1}{2m}\left(\frac{\partial\xi}{\partial x}\right)^2 + V + \frac{\partial\xi}{\partial t}\right]\psi = 0 , \tag{5.45}$$

from which we obtain the classical Hamilton–Jacobi equation

$$\frac{1}{2m}\left(\frac{\partial S}{\partial x}\right)^2 + V + \frac{\partial S}{\partial t} = 0 , \tag{5.46}$$

where we have replaced $\xi = \xi(x,t)$ by the classical action $S = S(x,t)$ using the properties (5.40–5.42). In principle, (5.46) is obtained from (5.45) for all solutions ψ of the Schrödinger equation compatible with our eigenvalue equation (5.43). It is worth emphasizing that our classical limiting process goes through without the need to evaluate the limit $\hbar \to 0$ explicitly for certain functions ψ because they are not in general analytical in this limit. In short, our method works well for any quantum isolated system described in terms of ψ, since (5.43) is satisfied.

It is straightforward to check that both solutions (5.35) of a free particle, and also their superposition

$$\psi(x,t) = \frac{2A}{\sqrt{2}} \mathrm{e}^{(-\mathrm{i}/2m\hbar)p^2} \cos\frac{px}{\hbar} , \tag{5.47}$$

satisfy the relation (5.43). As another example, let us consider a system described in the semiclassical domain of quantum mechanics where the potential function $V = V(x)$ has the behavior

$$\left|\frac{\mathrm{d}V}{\mathrm{d}x}\right| \approx 0 . \tag{5.48}$$

The WKB functions (5.31) in the form

$$\psi_{\mathrm{WKB}}^{(\pm)} = \mathrm{e}^{(\mathrm{i}/\hbar)[\pm S(x) - Et]} , \tag{5.49}$$

with $\partial^2 S/\partial x^2 = 0$, are good approximate solutions of the Schrödinger equation. Since each function in (5.49) and their superposition $\psi = \psi_{\mathrm{WKB}}^{(+)} + \psi_{\mathrm{WKB}}^{(-)}$, which is not itself a WKB function, obey the condition (5.43), the Hamilton–Jacobi equation (5.46) is obtained in the classical limit $\hbar \to 0$.

Note that, having obtained the classical equation (5.46), we can also evaluate the classical limit of quantum kinematics. For example, for the Born–Jordan–Dirac relation $[\hat{p}, \hat{x}] = \hat{p}\hat{x} - \hat{x}\hat{p} = -\mathrm{i}\hbar$, we find the result

$$\lim_{\hbar\to 0}\left\{\mathrm{e}^{-\mathrm{i}S/\hbar}[\hat{p}, \hat{x}]\mathrm{e}^{\mathrm{i}S/\hbar} = -\mathrm{i}\hbar\right\} \quad \Longrightarrow \quad px = xp . \tag{5.50}$$

Thus, using our classical limiting process, we are able to show how the non-commutative algebraic structure of quantum mechanics in the Schrödinger representation disappears in the classical limit [13].

Summarizing, according to our method (5.38–5.46), we have shown that the validity conditions of the WKB approximation are only sufficient but not necessary for evaluating the classical limit of the Schrödinger equation. In contrast with the Bohm approach, the superposition of functions like (5.47) is not excluded by our classical limiting process.

Classical Limit in Phase Space

Time Evolution of the Wigner Function

The Schrödinger equation is obtained from the von Neumann equation by means of the variable separation $\rho(x_1, x_2, t) = \psi(x_1, t)\psi^\dagger(x_2, t)$. The Fourier transform

$$W(p, x, t) = \frac{1}{2\pi} \int_{-\infty}^{\infty} \rho(x_1, x_2, t) \mathrm{e}^{-\mathrm{i}p\eta} \mathrm{d}\eta \,, \tag{5.51}$$

with

$$x_1 = x + \frac{\eta\hbar}{2} \,, \qquad x_2 = x - \frac{\eta\hbar}{2} \,, \tag{5.52}$$

gives rise to the Wigner representation of quantum mechanics [262].

We wish to evaluate the classical limit of the time evolution equation for the (quantum) Wigner function $W(p, x, t)$ (5.51) given by

$$\left[\hbar\frac{\partial}{\partial t} + \hbar\frac{p}{m}\frac{\partial}{\partial x} - \hbar\sum_{k=0}^{\infty} \frac{(-1)^k(\hbar/2)^{2k}}{(2k+1)!}\frac{\partial^{2k+1}V}{\partial x^{2k+1}}\frac{\partial^{2k+1}}{\partial p^{2k+1}}\right] W = 0 \,, \tag{5.53}$$

Taking $\hbar \to 0$ directly in (5.53) (divided by $\hbar$) to obtain the classical Liouville equation immediately is not generally a correct procedure, since W is indeed a quantum object, i.e., it involves $\hbar$ and does not have a well defined limit when $\hbar \to 0$. For potentials of the form $V = a + bx + cx^2$, the Wigner equation (5.53) reduces exactly to

$$\frac{\partial W}{\partial t} = -\frac{p}{m}\frac{\partial W}{\partial x} + \frac{\partial V}{\partial x}\frac{\partial W}{\partial p} \,. \tag{5.54}$$

Nevertheless, making the Wigner function W propagate classically does not mean that we obtain the classical limit of (5.53) [263].

By means of the transformation of phase space

$$W'(p, x, t) = \mathrm{e}^{-(\alpha/\hbar)\xi(p,x,t)} W(p, x, t) \tag{5.55}$$

and assuming the parameter α to be infinitesimal, i.e.,

$$\alpha^2 \approx 0 \tag{5.56}$$

in the limit $\hbar \to 0$, we obtain the classical Liouville equation for the probability distribution $\xi \equiv F(p,q,t)$:

$$\frac{\partial F}{\partial t} = -\frac{p}{m}\frac{\partial F}{\partial x} + \frac{\partial V}{\partial x}\frac{\partial F}{\partial p} , \tag{5.57}$$

provided that the following asymptotics exist:

$$\lim_{\hbar \to 0} W' \sim W'' \neq 0 , \tag{5.58}$$

$$\lim_{\hbar \to 0} \hbar^n W' \sim 0 \quad (n = 2, 4, 6, \ldots, \infty) , \tag{5.59}$$

$$\lim_{\hbar \to 0} \hbar \frac{\partial W'}{\partial q} \sim 0 \quad (q = x, t) , \tag{5.60}$$

$$\lim_{\hbar \to 0} \hbar^j \frac{\partial^n W'}{\partial p^n} \sim 0 \quad (j, n = 1, 2, 3, \ldots, \infty) . \tag{5.61}$$

In expression (5.61), $n \leq j$ for j even and $n = j$ for j odd. From (5.58), note that we obtain the classical limit of (5.53) without requiring the Wigner function to become equal to the classical probability density in the classical limit $\hbar \to 0$.

Time Evolution of the Kirkwood Function

It is well known that the phase space representation of quantum mechanics is not unique. Using the Kirkwood function [264]

$$K(p,x,t) = \frac{1}{2\pi}\int_{-\infty}^{\infty} \rho(x + \hbar\eta, x)\mathrm{e}^{-\mathrm{i}p\eta}\mathrm{d}\eta , \tag{5.62}$$

we find the equation of motion

$$\left[\mathrm{i}\hbar\frac{\partial}{\partial t} + \mathrm{i}\hbar\frac{p}{m}\frac{\partial}{\partial x} - \frac{\hbar^2}{2m}\frac{\partial^2}{\partial x^2} - \sum_{k=1}^{\infty}\frac{(\hbar\mathrm{i})^k}{k!}\frac{\partial^k V}{\partial x^k}\frac{\partial^k}{\partial p^k}\right] K = 0 . \tag{5.63}$$

Note that, due to the presence of $\mathrm{i} = \sqrt{-1}$ in (5.63), K is a complex-valued function. Now performing the transformation

$$K'(p,x,t) = \mathrm{e}^{(\mathrm{i}\alpha/\hbar)\xi(p,x,t)} K(p,x,t) \tag{5.64}$$

on (5.63) and using assumption (5.56), we arrive at the Liouville equation (5.57) with the asymptotic conditions

$$\lim_{\hbar\to 0} K' \sim K'' \neq 0 \,, \tag{5.65}$$

$$\lim_{\hbar\to 0} \hbar^n K' \sim 0 \quad (n = 1, 2, 3, \ldots, \infty) \,, \tag{5.66}$$

$$\lim_{\hbar\to 0} \hbar \frac{\partial K'}{\partial q} \sim 0 \quad (q = x, t) \,, \tag{5.67}$$

$$\lim_{\hbar\to 0} \hbar^j \frac{\partial^k K'}{\partial p^k} \sim 0 \quad (k, j = 1, 2, \ldots, \infty) \quad k \leq j \,. \tag{5.68}$$

Time Evolution of the Husimi Function

The equation of motion for the Husimi function

$$\mathcal{H}(p, x, t; \beta) = \frac{1}{2\pi} \int_{-\infty}^{\infty} \int_{-\infty}^{\infty} W(x', p', t) \mathrm{e}^{-(x'-x)^2/\beta} \mathrm{e}^{-\beta(p'-p)^2/\hbar^2} \mathrm{d}x' \mathrm{d}p' \,, \tag{5.69}$$

where $\beta = \hbar/m\omega$, is given by [265]

$$\begin{aligned} 0 = \hbar \frac{\partial \mathcal{H}}{\partial t} + \hbar \frac{p}{m} \frac{\partial \mathcal{H}}{\partial x} + \frac{\hbar^2}{2\gamma m} \frac{\partial^2 \mathcal{H}}{\partial p \partial x} \\ -\hbar \sum_{k\lambda\mu} \frac{(\mathrm{i}\hbar)^{\lambda-1} \gamma^{\mu-1}}{2^{\lambda+\mu-1} \lambda! k! (\mu - 2k)!} \frac{\partial^{\lambda+\mu} V}{\partial x^{\lambda+\mu}} \frac{\partial^\lambda}{\partial p^\lambda} \frac{\partial^{\mu-2k} \mathcal{H}}{\partial x^{\mu-2k}} \,. \end{aligned} \tag{5.70}$$

The infinitesimal transformation

$$\mathcal{H}'(p, x, t) = \mathrm{e}^{(\alpha/\hbar)\xi(p,x,t)} \mathcal{H}(p, x, t) \tag{5.71}$$

leads to the Liouville equation (5.57), too. This result is only conjectured by O'Connell and Wigner [265]. In (5.70), $\lambda = 1, 3, 5, \ldots, \infty$, $\mu = 0, 1, 2, \ldots, \infty$, $k = 0, 1, 2, \ldots, \infty$ since $\mu - 2k > 0$. The asymptotic conditions are

$$\lim_{\hbar\to 0} \mathcal{H}' \sim \mathcal{H}'' \neq 0 \,, \tag{5.72}$$

$$\lim_{\hbar\to 0} \hbar^n \mathcal{H}' \sim 0 \quad (n = 1, 2, 3, \ldots, \infty) \,, \tag{5.73}$$

$$\lim_{\hbar\to 0} \hbar^n \frac{\partial^{i+j} \mathcal{H}'}{\partial u^i \partial v^j} \sim 0 \quad (u, v = p, x;\; n, i, j = 1, 2, \ldots, \infty;\; n \geq i + j) \,, \tag{5.74}$$

$$\lim_{\hbar\to 0} \hbar^n \frac{\partial^j \mathcal{H}'}{\partial q^j} \sim 0 \quad (q = p, x; n, j = 1, 2, \ldots, \infty) \,. \tag{5.75}$$

In order to assert that the function ξ in the transformations (5.55), (5.64), and (5.71) really does have the properties of a probability density of classical statistical mechanics, we should use the following criterion: if we begin with some probability (quasi-)density function in quantum phase space $(x, p; \hbar)$ [e.g., the Wigner function solution of (5.53)], then we hope that physically we have a true density function $F(x, p, t)$ in classical phase space (q, p), solving the Liouville equation (5.57), neglecting other mathematically admissible functions.

An Example: The Einstein Problem

With the purpose of investigating the classical limit of quantum mechanics, Einstein [266] proposed the following problem: to solve the Schrödinger equation for a macroscopic particle with diameter 1 mm and constant energy constrained to moving between two reflecting walls 1 m apart. The external potential is null for $-a < x < a$ and infinite at points $x = -a$ and $x = a$. The boundary conditions are $\psi(x = -a, t) = 0$ and $\psi(x = a, t) = 0$. The solutions of this problem are given by

$$\psi_{n'}(x, t) = \left(\frac{2}{a}\right)^{1/2} \cos\left(\frac{n'\pi}{a}x\right) \mathrm{e}^{\mathrm{i}tE_{n'}/\hbar} , \qquad (5.76)$$

where

$$E_{n'} = \frac{1}{2m}\left(\frac{\hbar n'\pi}{a}\right)^2 , \quad n' = n + \frac{1}{2} \quad (n = 0, 1, 2, \ldots) .$$

Although this example is mathematically well established, several interpretations arise from different points of view.

Einstein's View

Assuming the universality of quantum mechanics, according to Einstein [266,267] the classical limit should exist for all Schrödinger functions ψ. According to the Born statistical interpretation, the particle moves with the same probability of having a positive momentum p (from left to right) and a negative one $-p$ (from right to left). However, in the classical limit a macroscopic particle is either moving with momentum p or with momentum $-p$. This classical domain is characterized physically by the condition that the de Broglie wavelength $\lambda = 2a/n'$ should be small compared with the distance $2a$, or equivalently by

$$2a\frac{\hbar n'\pi}{a} \gg 2\pi\hbar \quad \text{or formally} \quad \hbar \to 0\,, \tag{5.77}$$

meaning that the quantum of action is negligible compared with the action of a classical particle of momentum $p = \hbar n'\pi/a$. Thus, the function ψ is not associated with any ontological properties of matter, such as the trajectory concept. For Einstein this simple example shows that ψ yields an incomplete description of physical reality. It provides only statistical features. For an ensemble of say 1000 copies of the system, in 50% of the cases, the particle is found with momentum p and in 50% with momentum $-p$. A new theory is therefore needed to justify those statistical properties.

In summary, the universality of the Schrödinger equation and the Born interpretation do not correctly link the quantum theory to the classical theory based on individual systems, since ψ plays an epistemological role while classical physics is based on ontological characteristics. In other words, it is impossible to derive ontology from epistemology.

Pauli's View

In quantum mechanics the state of a system has to be specified by the Schrödinger function ψ as well as by the measuring apparatus. According to the universality of ψ, Einstein's macroscopic particle should present interference effects. Nevertheless, as these effects are neglected due to the smallness of Planck's constant, they cannot be detected experimentally. According to Pauli [268], the quantum laws treat only the statistics of the acts of observation: "The appearance of a definite position $[x_0]$ during a subsequent observation [...] is then regarded as being a 'creation' existing outside the laws of nature" [268]. In Pauli's view, physics is pure epistemology: there is no place for ontology within physics.

Born's View

Besides sharing Pauli's position as regards the elimination of ontology from physics, Born [269] solves Einstein's problem mathematically using the d'Alembert method and shows that the solution (5.76) is not adequate to study the classical limit of quantum mechanics. Assuming then the universality of ψ, the quantum mechanical probability density found by Born reduces correctly to the classical one in the limit $\hbar \to 0$.

Bohm's View

The quantum potential associated with (5.76) is given by [15,230,257]

$$Q_{n'} = \frac{1}{2m}\left(\frac{\hbar n'\pi}{a}\right)^2, \tag{5.78}$$

vanishing as $\hbar \to 0$ (with n' fixed). The wave function (5.76) predicts zero probability of finding the macroscopic particle at its nodal points (neglected in Einstein's analysis). However, in the classical domain, there are no nodes. In the Bohm theory, the solution (5.76) leads to the case in which the particle is always at rest ($p = \partial S/\partial q = 0$) everywhere except at the nodes where p is not defined. Although this result is consistent with the existence of nodes, the quantum potential (5.78) is different from zero in the limit $\hbar \to 0$ and $n' \to \infty$. Thus, in spite of attempts to obtain the classical limit in a correct way (e.g., by appealing to a measurement device), Einstein's simple example reveals that the Madelung representation of quantum mechanics is not a suitable locus for investigating the connection between quantum and classical physics. Consequently, $Q \to 0$ is not a mathematically universal criterion for calculating the classical limit of quantum mechanics.

Decoherence Approach

According to the decoherence approach (Sect. 5.6), the macroscopic particle considered by Einstein is indeed a quantum open system, that is, a system never isolated from environmental influences. Hence, friction and fluctuation effects need to be taken into account in order for the quantum-to-classical transition to occur. On the basis of the universality assumption, particle and environment form an isolated whole to be described by the Schrödinger equation. By eliminating the degrees of freedom of the environment, we obtain stochastic systems to be described by quantum master equations. Environmental influences are responsible for destroying quantum superpositions, introduced via initial conditions (decoherence process).

Our View

In contrast with the above views, our quantization (Sect. 4.9) and classical limiting methods reveal the non-universal nature of both quantum and classical mechanics. Thus, in accordance with Einstein, our process $\hbar \to 0$ is performed without making reference to any act of observation. It is also compatible with all solutions of the Schrödinger equation, since the eigenvalue equation (5.43) or the asymptotics (5.58–5.61) are obeyed. In contrast with Einstein and Bohm, the ontological character of quantum physics is guaranteed by the existence of our quantization method (Sect. 4.9), showing that the main quantum object is the von Neumann function.

In the case of Einstein's problem, the function $\psi = 0$ is excluded on physical grounds because it corresponds to nodal points which do not exist in the classical domain. Moreover, our classical limiting process shows that, in the classical limit (4.77), Einstein's particle must be described by the Hamilton–Jacobi equation or the Liouville equation. Here its momentum always has the value $p = \partial S/\partial x \neq 0$. This differs from the Bohm approach which predicts $p = 0$ and no classical limit.

If we consider a more realistic situation involving some dissipation effects, we obtain a description in terms of the von Neumann equation, not reducible to any Schrödinger equation. No decoherence process needs to be invoked to calculate the classical limit of the Caldeira–Leggett equation, for instance. This issue is addressed below.

5.8.2 Open Systems

Classical Limit of the Caldeira–Leggett Master Equation

Performing the Wigner transformation (5.51) on the Caldeira–Leggett equation (4.92), we find the following equation in quantum phase space:

$$\hbar\frac{\partial W}{\partial t} + \hbar\frac{p}{m}\frac{\partial W}{\partial x} - \hbar\frac{\partial V}{\partial x}\frac{\partial W}{\partial p} + \mathcal{G}W - \hbar 2\gamma W - \hbar 2\gamma p\frac{\partial W}{\partial p} - \hbar D\frac{\partial^2 W}{\partial p^2} = 0\,, \tag{5.79}$$

where

$$\mathcal{G}W = -\frac{2}{\mathrm{i}3!}\left(\frac{-\hbar}{2\mathrm{i}}\right)^3\frac{\partial^3 V}{\partial x^3}\frac{\partial^3 W}{\partial p^3} - \frac{2}{\mathrm{i}5!}\left(\frac{-\hbar}{2\mathrm{i}}\right)^5\frac{\partial^5 V}{\partial x^5}\frac{\partial^5 W}{\partial p^5} - \cdots\,. \tag{5.80}$$

The infinitesimal transformation (5.55) and the classical limit $\hbar \to 0$ lead to the equation of motion

$$\frac{\partial \xi}{\partial t} + \frac{p}{m}\frac{\partial \xi}{\partial x} - \frac{\partial V}{\partial x}\frac{\partial \xi}{\partial p} - 2\gamma\xi - 2\gamma p\frac{\partial \xi}{\partial p} - D\frac{\partial^2 \xi}{\partial p^2} = 0\,, \tag{5.81}$$

which is the Fokker–Planck equation for the probability density function $\xi = D_{XP}(x,p,t)$ [see (3.128)]. Equation (5.81) is obtained since the asymptotic conditions (5.58–5.61) and

$$\lim_{\hbar\to 0} \hbar W' \sim \alpha\xi W'' , \tag{5.82}$$

$$\lim_{\hbar\to 0} \frac{\partial W'}{\partial p} \sim 0 \tag{5.83}$$

are satisfied. We can then calculate the classical limit of (5.79) without taking into account the high-temperature condition. As we have already emphasized in Sect. 4.9, such a condition is an artifact of the Hamiltonian model assumed by Caldeira and Leggett.

In order to calculate the classical limit of (5.79), Caldeira and Leggett made use of an argument based on the WKB expansion expressed in terms of the Wigner function [130]: "in order to be consistent with the fact we are considering [the high-temperature case], we need to take the limit $\hbar \to 0$ in the expression for W [(5.79)]. When $\hbar \to 0$, $W(p, x, t)$ tends to the classical distribution in phase space. Then we conclude that [(5.79)] describes the time development of the phase space distribution of a classical Brownian particle when $\hbar \to 0$." If the quantum function W is expanded as

$$W(p, x, t) = D_{XP} + \hbar^2 g_1 + \hbar^4 g_2 + \cdots , \tag{5.84}$$

where $D_{XP}, g_1, g_2, \ldots$ are functions that are independent of $\hbar$, one easily arrives at (5.81). Physically, one can justify the expansion (5.84) only for high-temperature cases [263].

The novel feature present in our classical limiting process lies in the asymptotic conditions (5.58–5.61), (5.82) and (5.83). This shows that the Caldeira–Leggett argument is not general enough to obtain the classical limit of quantum Brownian dynamics.

Classical Limit of our Non-Markovian Master Equation

In terms of the Wigner function (5.51), our quantum master equation (4.105) given by

$$\begin{aligned} 0 = {} & \mathrm{i}\hbar\frac{\partial \rho}{\partial t} + \frac{\hbar^2}{2m}\left(\frac{\partial^2 \rho}{\partial x_1^2} - \frac{\partial^2 \rho}{\partial x_2^2}\right) - \Big[V(x_1, t) - V(x_2, t)\Big]\rho \\ & + \Big[\mathrm{i}\hbar A(t) - (x_1 - x_2)C(t)\Big]\left(\frac{\partial \rho}{\partial x_1} + \frac{\partial \rho}{\partial x_2}\right) \\ & + \frac{\mathrm{i}\hbar}{2}\gamma(x_1 - x_2)\left(\frac{\partial \rho}{\partial x_1} - \frac{\partial \rho}{\partial x_2}\right) - \mathrm{i}\hbar B(t)\left(\frac{\partial}{\partial x_1} + \frac{\partial}{\partial x_2}\right)^2 \rho \end{aligned} \tag{5.85}$$

reads

$$0 = \hbar\frac{\partial W}{\partial t} + \hbar\frac{p}{m}\frac{\partial W}{\partial x} - \hbar\frac{\partial V}{\partial x}\frac{\partial W}{\partial p} + \mathcal{G}W \tag{5.86}$$
$$+A(t)\hbar\frac{\partial W}{\partial x} - \hbar B(t)\frac{\partial^2 W}{\partial x^2} - \hbar C(t)\frac{\partial^2 W}{\partial x \partial p} - \hbar\gamma W - \hbar\gamma p\frac{\partial W}{\partial p} = 0\ ,$$

where $\mathcal{G}W$ is given by (5.80).

The classical limit $\hbar \to 0$ leads correctly to our non-Markovian Fokker–Planck equation [see (4.100)]

$$0=\frac{\partial\xi}{\partial t}+\frac{p}{m}\frac{\partial\xi}{\partial x}+A(t)\frac{\partial\xi}{\partial x}-\frac{\partial V}{\partial x}\frac{\partial\xi}{\partial p}-\gamma\xi-\gamma p\frac{\partial\xi}{\partial p}-B(t)\frac{\partial^2\xi}{\partial x^2}-C(t)\frac{\partial^2\xi}{\partial x\partial p}\ , \tag{5.87}$$

since we have the following asymptotic limits in quantum phase space:

$$\lim_{\hbar\to 0} W' \sim W'' \neq 0\ , \tag{5.88}$$

$$\lim_{\hbar\to 0} \hbar W' \sim \alpha\xi\ , \tag{5.89}$$

$$\lim_{\hbar\to 0} \hbar\frac{\partial W'}{\partial u} \sim 0 \quad (u = t, x)\ , \tag{5.90}$$

$$\lim_{\hbar\to 0} \hbar\frac{\partial^2 W'}{\partial x^2} \sim 0\ , \tag{5.91}$$

$$\lim_{\hbar\to 0} \hbar\frac{\partial^2 W'}{\partial x\partial p} \sim 0\ , \tag{5.92}$$

$$\lim_{\hbar\to 0} \frac{\partial W'}{\partial x} \sim 0\ , \tag{5.93}$$

$$\lim_{\hbar\to 0} \hbar^j\frac{\partial^n W'}{\partial p^n} \sim 0 \quad (j, n = 1, 2, 3, \ldots, \infty)\ . \tag{5.94}$$

In (5.94), $n \leq j$ for j even and $n = j$ for j odd.

In contrast with Ballentine's statement [270]: "In spite of some attractive formal properties of the Wigner representation, it does not seem to provide a good approach to the classical limit", our results support the view that the phase space representation is the most adequate locus for evaluating the classical limit of quantum equations of motion, since it can be applied to both isolated and non-isolated systems. (In [45,46], we have also extended our method for calculating the classical limit of fermions and bosons in relativistic quantum phase space.) Furthermore, in phase space our method exhibits the relevance of the infinitesimal nature of the transformations (5.55), (5.64), and (5.71) for connecting quantum physics to classical physics. However, we should point out that the physical reasons behind such an infinitesimality condition as well as its relationship with the Olavo–Einstein infinitesimality (4.93) remain to be revealed. Despite this, we could justify the physical consistency of our classical limiting process ($\hbar \to 0$) from a purely operational standpoint: it just works!

6 Summary and Open Questions

6.1 Summary

The main subject of our monograph has been the account of the relationship between quantum physics and classical physics. We have sought to answer in a general way the following questions:

- How can we obtain the quantum dynamics of open systems initially described by the equations of motion of classical physics (quantization process)?
- How can we retrieve classical dynamics from the quantum mechanical equations of motion by means of a classical limiting process (dequantization process)?

The guiding theme of our work has been the concept of open systems where probability plays an ontological role in the definition. Open systems are physical systems interacting with their environment (e.g., a non-Markovian, non-Gaussian, and nonlinear Brownian particle). The richness of this interaction can only be apprehended in probabilistic terms. Environmental influences are encapsulated in the friction and diffusion coefficients present in the stochastic dynamics (e.g., Langevin equations and Fokker–Planck equation in the classical domain or the master equations in the quantum domain). When the environment can be neglected, we obtain an isolated system as a special case to be described by the Liouville equation or the von Neumann equation.

The usual approach to open systems is based on the following tenets:

- Universality of the concept of isolated system. As a consequence of this assumption, all open systems should be reduced to the statistical physics of isolated systems.
- Hamiltonian models must be constructed from which both the quantum and classical dynamics can be derived:

Hamiltonian model

$\swarrow \qquad \searrow$

Langevin and Fokker–Planck equations | quantum master equations

Here one regards both the system and the environment as an isolated whole to be described by a Hamiltonian function. The environment is imagined as consisting of a set of harmonic oscillators independent of each other. The quantization procedure used is Dirac or canonical quantization. From this model, several techniques for deriving quantum master equations have been developed (e.g., the Feynman–Vernon influence functional, and the Lindblad axiomatic approach).

The important point is that all Hamiltonian formulations of open systems support the epistemological role played by the probability concept. Probability enters owing to the operational difficulties of specifying initial conditions. According to a non-positivist view (e.g., Einstein), all probabilistic features should be justified on the basis of a more fundamental physical theory of isolated systems, whereas according to a positivist attitude, it should be used in order to evaluate its practical efficacy and efficiency. Here no theoretical derivation process is required.

On the contrary, in our approach to open systems, chance turns out to be an intrinsic or ontological property of physics. The central point of our present survey has consisted in exploring the fertility and generality of the theory of open systems by investigating a new method of quantization independent of Hamiltonians (dynamical quantization). We have also defined an alternative method for calculating the classical limit ($\hbar \to 0$) of quantum dynamics. We summarize as follows.

We start with the Langevin equations and the Fokker–Planck equation describing open systems in phase space. By means of a Fourier transform, we introduce the Wigner representation of classical mechanics characterized by an arbitrary parameter ℓ with the dimensions of action. Quantization occurs when we take $\ell \to \hbar$ in the time evolution for the classical Wigner function χ. Thus, in the von Neumann representation, we obtain quantum master equations (e.g., the Caldeira–Leggett equation) without restricting to a Hamiltonian framework, and without making unnecessary assumptions concerning initial conditions, a high temperature condition, or the weakness of friction.

In order to address the problem of the classical limit or dequantization $\hbar \to 0$, we perform a Fourier transform on the von Neumann function (or density matrix) and we arrive at the phase space representation of quantum mechanics. Here we take the limit $\hbar \to 0$ of quantum dynamics, thus recovering the classical equations from which we started. In this way, we are able to close the circle as regards the relationship between quantum and classical physics. Physically, we must bear in mind that such a quantum–classical correspondence is indeed a quantization–dequantization process. We illustrate this process by the following scheme:

$$\begin{array}{c} \text{classical physics of open systems} \\ \Downarrow \\ \text{Wigner representation} \\ \Downarrow \\ \text{quantization} \quad (\ell \to \hbar) \\ \Downarrow \\ \text{quantum physics of open systems} \\ \Downarrow \\ \text{von Neumann representation} \\ \Downarrow \\ \text{phase space representation} \\ \Downarrow \\ \text{dequantization} \quad (\hbar \to 0) \\ \Downarrow \\ \text{classical physics of open systems} \end{array}$$

Below we wish to list some epistemological advantages as regards the utility of establishing a quantization method for open systems without using Hamiltonian models, as well as its ontological relevance.

Epistemological Advantages

How useful is it to establish a quantization method without any Hamiltonian or Lagrangian?

The first advantage concerns the physical aspects. The Markov property comes from the structure subjacent to the Langevin equations. It does not therefore emerge from the Hamiltonian framework, assumed by Caldeira and Leggett, for instance. Furthermore, there is a kind of economy of calculus: we do not need to elaborate a Hamiltonian model, as we start directly from the stochastic differential equation. In general we are thus able to obtain non-Markovian, nonlinear and non-Gaussian quantum master equations from our dynamical quantization scheme, thus avoiding the difficulties of manipulation and the artificial hypothesis introduced to make Feynman's path integral formalism work. Such a procedure may be useful in practical applications such as open systems in quantum optics [154,271], where one assumes that an atom interacts with a heat bath consisting of many harmonic oscillators representing the electromagnetic field, in condensed matter physics (e.g., Josephson junction) [136], and also in the dynamics of open nanostructures [272], where Hamiltonian models are used to describe quantum Brownian motion.

Our method of quantizing anomalous Brownian motion may be useful in investigating the effects of possible quantum mechanical interactions between a diffusing particle and impurities in disordered media [83].

Another advantage of quantizing without Hamiltonians is the fact that it is easier experimentally (e.g., in superconductor physics [255]) to establish the phenomenological character of the equations of motion than to construct Hamiltonian models aimed at reproducing such phenomenological equations.

Ontological Advantages

How relevant is it conceptually to establish a quantization method without using the Hamiltonian framework?

- According to dynamical quantization, quantum physics can be formulated without appealing in any way to a dualistic framework of matter (wave vs. particle). Consequently, the trajectory concept exists as an intrinsic or ontological attribute for both quantum and classical particles. An electron around an atomic nucleus, for example, actually follows a certain quantum trajectory, since the density matrix ρ does possess the same ontological status as the classical Wigner function χ. This characteristic is not destroyed by our quantization procedure.
- The existence of our quantization–dequantization process leads to novel ways of understanding the relationship between classical and quantum mechanics. From a logical point of view, neither form of mechanics has universal validity, that is, neither theory can be derived from the other. Both theories are necessary to describe certain realms of physical reality. Physically, we imagine the quantization–dequantization process as a kind of a domain-transition phenomenon governed by the Planck constant. Concep-

tually, there exists a circular relationship between quantum and classical physics. This circularity is only possible within the infinitesimality underlying the Olavo–Einstein condition (4.90) and our infinitesimal condition (5.56).

- The Schrödinger function ψ plays a secondary role in the theoretical framework of quantum physics. In general, open systems are not described in terms of ψ. The description in terms of the von Neumann function ρ is fundamental due to its logical priority and its physical generality.
- Our quantization–dequantization process contradicts the assumption of universality of the Hamiltonian structure. For example, we were able to quantize and calculate the classical limit of a non-Markovian Brownian particle without using Hamiltonians or Lagrangians, and calculate its classical limit without alluding to a decoherence process and without associating a Bohm quantum potential with this open system.

6.2 Open Questions

Let us now list some open questions arising from our work.

- To give a theoretical justification of the infinitesimal nature of the Olavo–Einstein condition (4.90) used to obtain the Fokker–Planck equation on the one hand and quantum master equations on the other.
- To give a theoretical justification of the infinitesimal character of the transformations (5.55), (5.64) and (5.71) used to get the classical limit of quantum dynamics in phase space.
- To perform the relativistic extension of dynamical quantization. (We have extended our classical limiting method to relativistic equations such as the Dirac and Duffin–Kemmer–Petiau equations, both in configuration space and in phase space [45,46].)
- To carry out physical applications of our quantum master equation (4.105) describing a nonlinear, non-Markovian, and non-Gaussian Brownian particle. A further issue is to confirm the utility of the quantization of anomalous Brownian motion achieved in Sect. 4.9.3.
- To explore in more physical detail the quantization–dequantization process as a phenomenon analogous to a phase transition.

To conclude, we would like to state that, at the beginning of the 20th century, Brownian motion was responsible for establishing an ontological basis for the atomic structure of matter, thus contradicting the positivist belief according to which atomicity was an unnecessary hypothesis. We hope our quantization–dequantization process of open systems will open the door to an ontological status for chance within physics. We emphasize that concepts such as irreversibility, synergeticity, chaos, and fractality can only be established ontologically within non-Hamiltonian physics. Hence we must bear in mind that the construction of a theory of open systems is still an open adventure ... *per ardua ad astra!*

A Elements of Set Theory

In this appendix we present the notions of set theory essential for understanding the probability concept. For more details, see [13,50,52].

A.1 Definitions

Set theory is based on three basic or primitive notions:

- the notion of set,
- the notion of element,
- the notion of pertinence or membership making sense of the ideas of set and element.

We denote sets by the letters $A, B, \Phi, \Psi, \ldots$, and their respective elements by $a, b, \phi, \psi, \ldots$. We use $a \in A$ to indicate that the object a is an element of A, and $a \notin A$ when a is not in A. In general, a set A is defined or specified by certain properties that its elements a possess. We thus write

$$A = \{a | \text{certain property of } a\} \ .$$

As an example, we mention the set $\mathbb{R}$ of real numbers whose elements r belong to the interval $r_1 < r < r_2$, i.e., $R = \{r | r_1 < r < r_2\}$.

We fix a set containing all elements of all the sets under consideration. This reference set is called the space $\Omega = \{\omega\}$. A class is a collection of sets in Ω. It is represented by the symbols $\mathcal{A}, \mathcal{B}, \mathcal{C}, \ldots$. The empty set $\emptyset$ contains no elements.

A finite set contains a finite number of elements, whereas an infinite set possesses an infinite number. An infinite set is said to be denumerable or countable if there exists a one-to-one correspondence between all its elements and all positive integers $A = \{1, 2, 3, \ldots\}$. If this condition is not satisfied, then the set is said to be non-denumerable or uncountable (e.g., the set of all points on a straight line segment).

If all the elements of a set A are contained in another set B, then A is a subset of B, written $A \subset B$ or $B \supset A$. The empty set $\emptyset$ is a subset of any set. Sets A and B are equal if both contain the same elements, i.e., $A = B$. If $A \subset B$ and $B \subset C$, then $A \subset C$.

The complement set of A, denoted by A^{c}, is the set of all elements in Ω which are not elements of A, i.e.,

$$A^{\mathrm{c}} = \{\omega | \omega \notin A, \omega \in \Omega\} \ .$$

The following relations hold:

$$(A^{\mathrm{c}})^{\mathrm{c}} = A \ , \quad \Omega^{\mathrm{c}} = \emptyset \ , \quad \emptyset^{\mathrm{c}} = \Omega \ .$$

A.2 Algebraic Structure

The algebraic structure in set theory is based on algebraic operations (addition, multiplication, and difference) involving sets $A, B, C, \ldots$. These operations generate new subsets belong to Ω.

- The union or sum of A and B, denoted by $A \cup B$, is the set of all $\omega \in \Omega$ such that ω is either an element of A or an element of B:

$$A \cup B = \{\omega | \omega \in A \text{ or } \omega \in B\} \ .$$

- The intersection or product, denoted by $A \cap B$ or AB, is the set of all elements belonging to both A and B:

$$A \cap B = \{\omega | \omega \in A \text{ and } \omega \in B\} \ .$$

- The difference between two sets, $(A - B)$, is the set of all elements belonging to A but not to B:

$$A - B = A \cap B^{\mathrm{c}} = \{\omega | \omega \in A \text{ and } \omega \notin B\} \ ,$$

  with the properties

$$A - \emptyset = A \ , \quad \Omega - A = A^{\mathrm{c}} \ , \quad A - B = AB^{\mathrm{c}} \ .$$

The sets A and B are said to be disjoint or mutually exclusive if $A \cap B = \emptyset$. The operations of union and intersection can be generalized to a countably infinite number of sets:

$$A_1 \cap A_2 \cap \ldots = \bigcap_{j=1}^{\infty} A_j \ .$$

The symbol $\bigcap_{j=1}^{\infty} A_j$ stands for the set of elements belonging to all the sets A_j, $j = 1, 2, \ldots$. The union

$$A_1 A_2 \ldots = \bigcup_{j=1}^{\infty} A_j$$

is the set of elements belonging to at least one of the sets A_j. The sets A_j are said to be mutually independent if $A_i A_j = \emptyset$, with $i \neq j$ $(j = 1, 2, \ldots)$.

The algebraic properties of addition and multiplication are:

- associativity: $(A \cup B) \cup C = A \cup (B \cup C)$ and $(A \cap B) \cap C = A \cap (B \cap C)$,
- commutativity: $A \cup B = B \cup A$ and $A \cap B = B \cap A$,
- distributivity: $A \cap (B \cup C) = (A \cap B) \cup (A \cap C)$ and $A \cup (B \cap C) = (A \cup B) \cap (A \cup C)$.

The following relations hold:

$$A \cup A = A \cap A = A \,,$$

$$A \cup \emptyset = A \,, \quad A \cap \emptyset = \emptyset \,,$$

$$A \cup \Omega = \Omega \,, \quad A \cap \Omega = A \,,$$

$$A \cup A^{\mathrm{c}} = \Omega \,, \quad A \cap A^{\mathrm{c}} = \emptyset \,,$$

$$(A \cap B)^{\mathrm{c}} = A^{\mathrm{c}} \cup B^{\mathrm{c}} \,, \quad (A \cup B)^{\mathrm{c}} = A^{\mathrm{c}} \cap B^{\mathrm{c}} \,.$$

De Morgan's law states that

$$(A_1 \cup A_2 \cup \ldots)^{\mathrm{c}} = A_1^{\mathrm{c}} \cap A_2^{\mathrm{c}} \cap \ldots \,,$$

$$(A_1 \cap A_2 \cap \ldots)^{\mathrm{c}} = A_1^{\mathrm{c}} \cup A_2^{\mathrm{c}} \cup \ldots \,.$$

The Cartesian product of the n sets $A_1, A_2, \ldots, A_n$ is the set

$$A_1 \times A_2 \times \ldots \times A_n = \{(\omega_1, \omega_2, \ldots, \omega_n) | \omega_i \in A_i, i = 1, 2, \ldots, n\} \,.$$

When $A_i = A$, $i = 1, 2, \ldots, n$, we denote the Cartesian product by A^n.

A.3 Borel Field

Given a space Ω and its subsets A_j, $j = 1, 2, \ldots$, assumed countable, a Borel field $\mathcal{B}$ is a class with the properties:

- $\Omega \in \mathcal{B}$,
- if $A_1 \in \mathcal{B}$, then $A_1^c \in \mathcal{B}$,
- if $A_j \in \mathcal{B}$, $j = 1, 2, \ldots$, then $\bigcup_{j=1}^{\infty} A_j \in \mathcal{B}$.

From these three properties, it follows that

$$\emptyset = \Omega^c \in \mathcal{B}$$

and

$$\bigcap_{j=1}^{\infty} A_j = \left(\bigcup_{j=1}^{\infty} A_j^c \right)^c \in \mathcal{B} \,.$$

Moreover, if $A \in \mathcal{B}$ and $B \in \mathcal{B}$, then $A - B = A \cap B^c \in \mathcal{B}$.

A Borel field $\mathcal{B}$ is therefore a class of sets (including $\emptyset$ and Ω) which is closed under all algebraic operations involving union, intersection and difference.

A.4 Set Function and Point Function

Let $\mathcal{A}$ be a class of subsets of Ω and $\mathbb{R}$ the set of real numbers (the real line going from $-\infty$ to $+\infty$). A set function f is a certain rule connecting or mapping a determined member of $\mathcal{A}$ to the set $\mathbb{R}$

$$f : \mathcal{A} \longrightarrow \mathbb{R} \,.$$

The domain and codomain of f are $\mathcal{A}$ and $\mathbb{R}$, respectively. We use the notation $f = f(A)$, where $f(A) \in \mathbb{R}$, i.e., $-\infty < f(A) < +\infty$, and $A \in \mathcal{A}$. As an example of a set function, we have the probability function $P = P(A)$ introduced in Chap. 2.

The set function $f(A)$ is said to be increasing if $f(A) \leq f(B)$ whenever $A \subset B$. It is said to be additive if, for any two disjoint sets $A, B \in \mathcal{A}$, we have

$$f(A \cup B) = f(A) + f(B) \,.$$

It is then easy to show that, for any three sets $A, B, C \in \mathcal{A}$, we have

$$f(A \cup B) = f(A) + f(B) - f(AB) \,,$$

$$f(A \cup B \cup C) = f(A) + f(B) + f(C) - f(AB) - f(AC) - f(BC) + f(ABC) \,.$$

When there exists a rule connecting each element $\omega \in \Omega$ to numbers in $\mathbb{R}$,

$$g : \Omega \longmapsto \mathbb{R} ,$$

the function $g = g(\omega)$ is said to be a point function, with $-\infty < g(\omega) < +\infty$. As an example of a point function, we can quote the stochastic variable $\Phi = \Phi(\omega)$ studied in Chap. 2.

B Quantization of the Smoluchowski Equation

In this appendix we shall quantize the Smoluchowski equation (3.31),

$$\frac{\partial f}{\partial t} = -\frac{1}{m\gamma}\frac{\partial}{\partial x}\Big[\mathcal{K}(x,t)f(x,t)\Big] + C\frac{\partial^2}{\partial x^2}f(x,t) \; . \tag{B.1}$$

B.1 Constant Force

Let us consider a Brownian particle described by the Smoluchowski equation (B.1) and subject to a constant force $\mathcal{K} = c$ so that

$$\frac{\partial f}{\partial t} = -\alpha\frac{\partial f}{\partial x} + C\frac{\partial^2 f}{\partial x^2} \; , \tag{B.2}$$

where $\alpha = c/m\gamma$. From (B.2), we derive

$$\frac{\partial \xi}{\partial t} = -\alpha\left(\frac{\partial \xi}{\partial x_1} + \frac{\partial \xi}{\partial x_2}\right) + C\left(\frac{\partial^2 \xi}{\partial x_1^2} + \frac{\partial^2 \xi}{\partial x_2^2}\right) \; , \tag{B.3}$$

where $\xi = \xi(x_1, x_2, t) = f(x_1,t)f(x_2,t)$. We make the variable change

$$x = \frac{x_1 + x_2}{\sqrt{2}} \; , \qquad \zeta = (x_1 - x_2)\sqrt{2} \; ,$$

and perform the transformation

$$F(x,p,t) = \frac{1}{2\pi\hbar}\int_{-\infty}^{\infty}\xi(x,\zeta,t)\mathrm{e}^{\mathrm{i}(p-p_0)\zeta/\hbar}\mathrm{d}\zeta \; .$$

The resulting equation is

$$\frac{\partial F}{\partial t} = -\sqrt{2}\alpha\frac{\partial F}{\partial x} + C\frac{\partial^2 F}{\partial x^2} - \frac{4C}{\hbar^2}(p-p_0)^2 F \; . \tag{B.4}$$

With the initial condition

$$F(x,p,t_0) = \frac{1}{\pi\hbar}\mathrm{e}^{-[a(p-p_0)^2 + b(x-x_0)^2]/\hbar} \quad (ab = 1) \; , \tag{B.5}$$

this has solution

$$F(x,p,t) = \frac{1}{\pi\hbar}\exp\left\{-\mathcal{A}(p-p_0)^2 - \frac{1}{4\mathcal{B}}\left[x - x_0 - \sqrt{2}\alpha(t-t_0)\right]^2\right\}, \tag{B.6}$$

where $\mathcal{A}$ and $\mathcal{B}$ are given by

$$\mathcal{A} = \frac{a}{\hbar} + \frac{4C}{\hbar^2}(t-t_0), \tag{B.7}$$

$$\mathcal{B} = \frac{\hbar}{4b} + C(t-t_0). \tag{B.8}$$

Equation (B.6) leads to

$$\langle (x-x_0)^2 \rangle = \frac{\hbar}{2b} + 2C(t-t_0), \tag{B.9}$$

$$\langle (p-p_0)^2 \rangle = \frac{\hbar^2}{2[a\hbar + 4C(t-t_0)]}, \tag{B.10}$$

$$\langle (x-x_0)^2 (p-p_0)^2 \rangle = \frac{\hbar^2}{4}. \tag{B.11}$$

B.2 Linear Force

Another example we need to consider is a Brownian particle under a linear force $\mathcal{K} = -kx$. Equation (B.1) reads

$$\frac{\partial f}{\partial t} = \theta\frac{\partial(xf)}{\partial x} + C\frac{\partial^2 f}{\partial x^2}, \tag{B.12}$$

where $\theta = k/m\gamma$. We begin from (B.12) at two different points x_1 and x_2. Multiplying the first equation by $f(x_2,t)$ and the second by $f(x_1,t)$, we arrive at

$$\frac{\partial\xi}{\partial t} = 2\theta\xi + \theta x_1\frac{\partial\xi}{\partial x_1} + \theta x_2\frac{\partial\xi}{\partial x_2} + C\left(\frac{\partial^2\xi}{\partial x_1^2} + \frac{\partial^2\xi}{\partial x_2^2}\right). \tag{B.13}$$

Following the same procedure as above, we obtain the quantum Smoluchowski equation

$$\frac{\partial F}{\partial t} = \theta F + \theta x\frac{\partial F}{\partial x} + C\frac{\partial^2 F}{\partial x^2} - \frac{\theta}{\hbar}(p-p_0)\frac{\partial F}{\partial p} - \frac{4C}{\hbar^2}(p-p_0)^2 F. \tag{B.14}$$

For the initial condition (B.5), this has solution

$$F(x,p,t) = \frac{1}{2\pi}\left(\frac{A}{B}\right)^{1/2} \mathrm{e}^{-A(p-p_0)^2}\mathrm{e}^{-(x-H)^2/4B}\,, \tag{B.15}$$

where

$$A = \left(\frac{a}{\hbar} - \frac{2C}{\hbar}\right)\mathrm{e}^{-2\beta(t-t_0)} + \frac{2C}{\hbar}\,, \tag{B.16}$$

$$B = \left(\frac{\hbar}{4b} - \frac{C}{2\beta}\right)\mathrm{e}^{-2\beta(t-t_0)} + \frac{C}{2\beta}\,, \tag{B.17}$$

$$H = x_0\mathrm{e}^{-\beta(t-t_0)}\,. \tag{B.18}$$

The solution (B.15) yields

$$\sigma_x^2 = 2B\,, \tag{B.19}$$

$$\sigma_{(p-p_0)}^2 = \frac{1}{2A}\,, \tag{B.20}$$

$$\sigma_x^2\sigma_{(p-p_0)}^2 = \frac{B}{A}\,, \tag{B.21}$$

which reduce to

$$\sigma_x^2 = \frac{C}{\theta}\left[1 - \mathrm{e}^{-2\theta(t-t_0)}\right]\,, \tag{B.22}$$

$$\sigma_{(p-p_0)}^2 = 0\,, \tag{B.23}$$

$$\sigma_x^2\sigma_{(p-p_0)}^2 = 0\,, \tag{B.24}$$

in the classical limit $\hbar \to 0$. In contrast to (B.11), for a quantum Brownian particle submitted to a linear force, the Heisenberg relation (B.21) does not remain constant in time. For $t \to \infty$, (B.19–B.21) become

$$(\sigma_x)^2 = \frac{C}{\theta}\,, \tag{B.25}$$

$$\sigma_{(p-p_0)}^2 = \frac{\hbar}{4C}\,, \tag{B.26}$$

$$\sigma_x^2\sigma_{(p-p_0)}^2 = \frac{\hbar}{4\theta}\,, \tag{B.27}$$

corresponding to the stationary solution of (B.14).

B.3 General Case

From the Smoluchowski equation (B.1), at points x_1 and x_2, we derive

$$\frac{\partial \xi}{\partial t} = -g\left(\frac{\partial \mathcal{K}_1}{\partial x_1} + \frac{\partial \mathcal{K}_2}{\partial x_2}\right)\xi - g\left(\mathcal{K}_1 \frac{\partial \xi}{\partial x_1} + \mathcal{K}_2 \frac{\partial \xi}{\partial x_2}\right) + C\left(\frac{\partial^2 \xi}{\partial x_1^2} + \frac{\partial^2 \xi}{\partial x_2^2}\right), \tag{B.28}$$

where $g = 1/m\gamma$, $\xi = f(x_1, t)f(x_2, t)$ and $\mathcal{K}_i = \mathcal{K}(x_i, t)$, $i = 1, 2$. Changing the variables and using the expansions

$$\begin{aligned}\mathcal{K}\left(\frac{\sqrt{2}}{2}x \pm \frac{1}{2\sqrt{2}}\zeta\right) &= \mathcal{K}\left(\frac{\sqrt{2}}{2}x\right) \pm \frac{\zeta}{2\sqrt{2}}\frac{\partial \mathcal{K}}{\partial x} + \frac{1}{2!}\left(\frac{\zeta}{2\sqrt{2}}\right)^2 \frac{\partial^2 \mathcal{K}}{\partial x^2} \\ &\quad \pm \frac{1}{3!}\left(\frac{\zeta}{2\sqrt{2}}\right)^3 \frac{\partial^3 \mathcal{K}}{\partial x^3} + \cdots \\ &= \sum_{n=0}^{\infty} (\pm)^n \frac{1}{n!}\left(\frac{\zeta}{2\sqrt{2}}\right)^n \frac{\partial^n \mathcal{K}}{\partial x^n},\end{aligned} \tag{B.29}$$

equation (B.28) becomes

$$\frac{\partial \xi}{\partial t} = -g\left(\mathcal{O}\xi + \mathcal{A}\frac{\partial \xi}{\partial x} + \mathcal{B}\frac{\partial \xi}{\partial \zeta}\right) + C\left(\frac{\partial^2 \xi}{\partial x^2} + 4\frac{\partial^2 \xi}{\partial \zeta^2}\right), \tag{B.30}$$

where

$$\mathcal{O} = \mathcal{O}(x, \zeta) = 3\sqrt{2} \sum_{k=1,3,5,\ldots} \frac{1}{(k-1)!}\left(\frac{\zeta}{2\sqrt{2}}\right)^{k-1} \frac{\partial^k \mathcal{K}}{\partial x^k}, \tag{B.31}$$

$$\mathcal{A} = \mathcal{A}(x, \zeta) = \frac{2}{\sqrt{2}} \sum_{r=0,2,4,\ldots} \frac{1}{r!}\left(\frac{\zeta}{2\sqrt{2}}\right)^r \frac{\partial^r \mathcal{K}}{\partial x^r}, \tag{B.32}$$

$$\mathcal{B} = \mathcal{B}(x, \zeta) = 2\sqrt{2} \sum_{k=1,3,5,\ldots} \frac{1}{k!}\left(\frac{\zeta}{2\sqrt{2}}\right)^k \frac{\partial^k \mathcal{K}}{\partial x^k}. \tag{B.33}$$

If we quantize (B.30), we obtain the quantum Smoluchowski equation

$$\frac{\partial F}{\partial t} = -g\left(\mathcal{O}'F + \mathcal{A}'F + \mathcal{B}'F\right) + C\frac{\partial^2 F}{\partial x^2} - \frac{4C}{\hbar^2}(p - p_0)^2 F, \tag{B.34}$$

where

$$\mathcal{O}'F = 3\sqrt{2} \sum_{k=1,3,5,\ldots} \frac{1}{(k-1)!(2\mathrm{i})^{k-1}} \frac{\partial^k \mathcal{K}}{\partial x^k} \frac{\partial^{k-1} F}{\partial p^{k-1}}, \tag{B.35}$$

$$\mathcal{A}'F = \frac{2}{\sqrt{2}} \sum_{r=0,2,4,\ldots} \frac{1}{r!2^r} \frac{\partial^r \mathcal{K}}{\partial x^r} \frac{\partial}{\partial x} \frac{\partial^r F}{\partial p^r} , \tag{B.36}$$

$$\mathcal{B}'F = 2\sqrt{2} \sum_{k=1,3,5,\ldots} \frac{1}{k!2^k \mathrm{i}^{k+1}} \frac{\partial^k \mathcal{K}}{\partial x^k} \left[k \frac{\partial^{k-1} F}{\partial p^{k-1}} + \frac{(p-p_0)}{\hbar} \frac{\partial^k F}{\partial p^k} \right] . \tag{B.37}$$

C Dynamical Quantization Versus Dirac Quantization

In this appendix, our objective is to compare our quantization method, which we have called dynamical quantization, with Dirac quantization.

In Chap. 3, we saw that the canonical rules

$$p \Longrightarrow \hat{p} = -\mathrm{i}\hbar \frac{\partial}{\partial x} \,, \tag{C.1}$$

$$x \Longrightarrow \hat{x} = x \,, \tag{C.2}$$

$$E \Longrightarrow \hat{E} = \mathrm{i}\hbar \frac{\partial}{\partial t} \,, \tag{C.3}$$

work as a very good quantization procedure since the Hamiltonian function is identified with the total energy of the system, i.e.,

$$H(p,x) = \frac{p^2}{2m} + V(x) = E \,. \tag{C.4}$$

It is straightforward to show that Dirac's criterion is not necessary to quantize a given classical system [43]. Let us consider the Hamiltonian [240]

$$H(p,x) = \sqrt{V(x)} \cosh\left(p\sqrt{\frac{2}{m}}\right) \,, \tag{C.5}$$

reproducing the same Newtonian equations as the Hamiltonian function (C.4), viz.,

$$\dot{p} = -\frac{\partial V}{\partial x} \,, \qquad \dot{x} = \frac{p}{m} \,, \tag{C.6}$$

which in turn generate the Liouville equation

$$\frac{\partial}{\partial t} D_{XP} + \dot{x}\frac{\partial}{\partial x} D_{XP} + \dot{p}\frac{\partial}{\partial p} D_{XP} = 0 \,. \tag{C.7}$$

As our quantization method starts directly from the Newtonian equations, through the Liouville equation (C.7), we do not need to identify the Hamiltonian with the total energy of the system (C.4).

On the other hand, Podolski [167] has already shown that the Dirac criterion (C.4) is also not sufficient. The transformation from (x, y, p_x, p_y) to the polar coordinates $(r, \theta, p_r, p_\theta)$ changes

$$H(x, y, p_x, p_y) = \frac{p_x^2 + p_y^2}{2m} + V(x, y) = E \tag{C.8}$$

to the Hamiltonian

$$H'(r, \theta, p_r, p_\theta) = \frac{p_r^2}{2m} + \frac{p_\theta^2}{2mr^2} + V(r) = E\ , \tag{C.9}$$

which is also identical with the total energy of the system. Consequently, we can quantize (C.9) using the Dirac rules (C.1–C.3) in the form

$$p_r \Longrightarrow \hat{p}_r = -\mathrm{i}\hbar\frac{\partial}{\partial r}\ , \tag{C.10}$$

$$p_\theta \Longrightarrow \hat{p}_\theta = -\mathrm{i}\hbar\frac{\partial}{\partial \theta}\ , \tag{C.11}$$

$$r \Longrightarrow \hat{r} = r\ , \tag{C.12}$$

$$E \Longrightarrow \hat{E} = \mathrm{i}\hbar\frac{\partial}{\partial t}\ . \tag{C.13}$$

However, this procedure is not correct, since the resulting equation

$$-\frac{\hbar^2}{2m}\frac{\partial^2\psi}{\partial r^2} - \frac{\hbar^2}{2mr^2}\frac{\partial^2\psi}{\partial\theta^2} + V\psi = \mathrm{i}\hbar\frac{\partial\psi}{\partial t} \tag{C.14}$$

is physically different from the true Schrödinger equation in polar coordinates, viz.,

$$-\frac{\hbar^2}{2m}\left(\frac{\partial^2\psi}{\partial r^2} + \frac{1}{r}\frac{\partial\psi}{\partial r} + \frac{1}{r^2}\frac{\partial^2\psi}{\partial\theta^2}\right) + V\psi = \mathrm{i}\hbar\frac{\partial\psi}{\partial t}\ . \tag{C.15}$$

Therefore, Dirac quantization only works in Cartesian coordinates. In contrast, using dynamical quantization, it is possible to quantize in any coordinate system [209].

In summary, the Dirac criterion (C.4) is neither necessary nor sufficient to quantize a given classical system correctly. Thus, there is no theoretical justification for using the canonical rules of quantization. The procedure (C.1–C.3) is entirely arbitrary.

D Dynamical Quantization Versus Feynman Quantization

On the basis of our work [44], we now compare dynamical quantization with Feynman quantization. We use the following transformation in phase space:

$$Q = e^{\beta(t)} x \,, \qquad P = e^{-\beta(t)} p - \alpha(t) e^{\beta(t)} x \,, \tag{D.1}$$

recently employed by Yeon et al. [191]. The terms $\beta(t)$ and $\alpha(t)$ are arbitrary functions of time t. Under the transformation (D.1), the Newtonian equations

$$\dot{p} = -\frac{\partial}{\partial x} V(x,t) \,, \qquad \dot{x} = \frac{p}{m} \tag{D.2}$$

can be written in terms of Q and P as

$$\dot{P} = \left(\dot{\beta} + e^{2\beta}\frac{\alpha}{m}\right) P - \left(2\alpha\dot{\beta} + \dot{\alpha} + e^{2\beta}\frac{\alpha^2}{m}\right) Q - \frac{\partial}{\partial Q} V \,, \tag{D.3}$$

$$\dot{Q} = e^{2\beta}\frac{P}{2m} + \left(\dot{\beta} + e^{2\beta}\frac{\alpha}{m}\right) Q \,. \tag{D.4}$$

These lead to the Liouville equation

$$\frac{\partial}{\partial t} F' + \dot{Q}\frac{\partial}{\partial Q} F' + \dot{P}\frac{\partial}{\partial P} F' = 0 \,, \tag{D.5}$$

where

$$F' = F'(P,Q,t) = F(p,x,t) \,. \tag{D.6}$$

We use the Fourier transform

$$\chi'(Q,\lambda,t) = \int_{-\infty}^{\infty} F'(P,Q,t) e^{iP\lambda} dP \,, \tag{D.7}$$

making the variable change

$$Q_1 = Q + \frac{\lambda\ell}{2} \,, \qquad Q_2 = Q - \frac{\lambda\ell}{2} \tag{D.8}$$

to obtain the equation of motion

$$\mathrm{i}\ell\frac{\partial\chi'}{\partial t}+\frac{\ell^2}{2m}\mathrm{e}^{2\beta}\left(\frac{\partial^2\chi'}{\partial Q_1^2}-\frac{\partial^2\chi'}{\partial Q_2^2}\right)-\Big[V(Q_1,t)-V(Q_2,t)+O'\Big]\chi'+\mathcal{B}\chi'=0\,, \tag{D.9}$$

where

$$O'(Q_1,Q_2,t)=-\sum_{n=3,5,7,\ldots}^{\infty}\frac{2^{1-n}}{n!}(Q_1-Q_2)^n\left(\frac{\partial}{\partial Q_1}+\frac{\partial}{\partial Q_2}\right)^n V(Q_1,Q_2,t) \tag{D.10}$$

and

$$\mathcal{B}=\mathrm{i}\ell\left(\dot\beta+\mathrm{e}^{2\beta}\frac{\alpha}{m}\right)\left(1+Q_1\frac{\partial}{\partial Q_1}+Q_2\frac{\partial}{\partial Q_2}\right) \\ -\frac{1}{2}\left(\mathrm{e}^{2\beta}\frac{\alpha^2}{m}+2\alpha\dot\beta+\dot\alpha\right)(Q_1^2-Q_2^2)\,. \tag{D.11}$$

We quantize (D.9) by taking into account the infinitesimality condition $(Q_1-Q_2)^3\to 0$ and the quantum limit $\ell\to\hbar$. The equation resulting from this procedure is

$$\begin{aligned}0=&\,\mathrm{i}\hbar\frac{\partial\rho'}{\partial t}+\frac{\hbar^2}{2m}\mathrm{e}^{2\beta}\left(\frac{\partial^2\rho'}{\partial Q_1^2}-\frac{\partial^2\rho'}{\partial Q_2^2}\right)-\Big[V(Q_1,t)-V(Q_2,t)\Big]\rho'\\ &+\mathrm{i}\hbar\left(\dot\beta+\mathrm{e}^{2\beta}\frac{\alpha}{m}\right)\rho'+\mathrm{i}\hbar\left(\dot\beta+\mathrm{e}^{2\beta}\frac{\alpha}{m}\right)\left(Q_1\frac{\partial\rho'}{\partial Q_1}+Q_2\frac{\partial\rho'}{\partial Q_2}\right)\\ &-\frac{1}{2}\left(\mathrm{e}^{2\beta}\frac{\alpha^2}{m}+2\alpha\dot\beta+\dot\alpha\right)(Q_1^2-Q_2^2)\rho'\,.\end{aligned} \tag{D.12}$$

Separating the variables according to $\rho'=\Psi^\dagger(Q_2,t)\Psi(Q_1,t)$, (D.12) implies

$$\begin{aligned}0=&\,\mathrm{i}\hbar\frac{\partial\Psi}{\partial t}+\frac{\hbar^2}{2m}\mathrm{e}^{2\beta}\frac{\partial^2\Psi}{\partial Q^2}+\frac{\mathrm{i}\hbar}{2}\left(\dot\beta+\mathrm{e}^{2\beta}\frac{\alpha}{m}\right)\Psi+\mathrm{i}\hbar\left(\dot\beta+\mathrm{e}^{2\beta}\frac{\alpha}{m}\right)Q\frac{\partial\Psi}{\partial Q}\\ &-\frac{1}{2}\left(\mathrm{e}^{2\beta}\frac{\alpha^2}{m}+2\alpha\dot\beta+\dot\alpha\right)Q^2-V(Q,t)\Psi\end{aligned} \tag{D.13}$$

at a generic point Q. This is the equation found by Yeon et al. [191] using the Lagrangian and Hamiltonian formalisms. However, in our derivation the ambiguities and inconsistencies inherent in the Feynman quantization method, such as the choice of midpoint, are avoided.

E Wigner Representation of Classical Mechanics

Recently [40–43], we have introduced the Wigner representation of classical mechanics. The relevance of this new representation is that it sets up a suitable route for quantizing dynamics as well as the operator structure of classical physics.

E.1 Dynamics

Let us consider a nonconservative deterministic system described by the Newtonian equations

$$\dot{p} = g(x,p,t)\ , \qquad \dot{x} = \frac{p}{m}\ . \tag{E.1}$$

The corresponding Liouville equation for the probability density function $F = F(x,p,t)$ is given by

$$\frac{\partial}{\partial t}F + \frac{p}{m}\frac{\partial}{\partial x}F = -\frac{\partial}{\partial p}(gF)\ . \tag{E.2}$$

Under the Fourier transform

$$\chi(x,\eta) = \int_{-\infty}^{\infty} F(x,p,t)\mathrm{e}^{\mathrm{i}p\eta}\mathrm{d}p\ , \tag{E.3}$$

the Liouville equation (E.2) reads

$$\mathrm{i}\ell\frac{\partial\chi}{\partial t} + \frac{\ell^2}{2m}\left(\frac{\partial^2\chi}{\partial x_1^2} - \frac{\partial^2\chi}{\partial x_2^2}\right) = -\mathrm{i}\ell\int_{-\infty}^{\infty}\frac{\partial}{\partial p}(gF)\mathrm{e}^{\mathrm{i}p\eta}\mathrm{d}p\ , \tag{E.4}$$

with

$$x_1 = x + \frac{\ell\eta}{2}\ , \qquad x_2 = x - \frac{\ell\eta}{2}\ . \tag{E.5}$$

Equation (E.4) determines the dynamics in the Wigner representation of classical mechanics. In Chap. 4, we showed that the quantization of our system occurs as $\ell \to \hbar$ and taking into account the infinitesimality condition $(x_1 - x_2)^3 \to 0$ in (E.4).

E.2 Kinematics

From (E.4), we note that the Wigner function χ is a complex-valued function due to the presence of $\mathrm{i} = \sqrt{-1}$. Using the inverse of (E.3), the mean value of a physical quantity $A(x, p, t)$ turns out to be

$$\langle A \rangle = \frac{1}{2\pi} \int_{-\infty}^{\infty} \int_{-\infty}^{\infty} \int_{-\infty}^{\infty} \chi(x, \eta, t) A(x, p, t) \mathrm{e}^{-\mathrm{i}p\eta} \mathrm{d}\eta \mathrm{d}p \mathrm{d}x \,. \tag{E.6}$$

The Wigner function has the following properties:

- hermiticity: $\langle \chi \rangle = \langle \chi^{\dagger} \rangle \Rightarrow \chi = \chi^{\dagger}$,
- non-positivity: even though $F \geq 0$, χ is not positive-definite.
- non-normalization: $\int_{-\infty}^{\infty} F \mathrm{d}x \mathrm{d}p = 1 \Rightarrow \int_{-\infty}^{\infty} \chi(x, x, t) \mathrm{d}x = 1$. However, $\chi(x_1, x_2, t)$ is not normalizable.

The Fourier transform induces an operator structure for classical mechanics. We use (E.6) for p:

$$\begin{aligned} \langle p \rangle &= \frac{1}{2\pi} \int_{-\infty}^{\infty} \int_{-\infty}^{\infty} \int_{-\infty}^{\infty} (p\chi) \mathrm{e}^{-\mathrm{i}p\eta} \mathrm{d}\eta \mathrm{d}p \mathrm{d}x \\ &= \frac{1}{2\pi} \int_{-\infty}^{\infty} \int_{-\infty}^{\infty} \int_{-\infty}^{\infty} \left(\mathrm{i} \frac{\partial}{\partial \eta} \chi \right) \mathrm{e}^{-\mathrm{i}p\eta} \mathrm{d}\eta \mathrm{d}p \mathrm{d}x \,. \end{aligned} \tag{E.7}$$

On the basis of (E.7), we can associate the Hermitian differential operator

$$\tilde{p} = \mathrm{i} \frac{\partial}{\partial \eta} \tag{E.8}$$

with the momentum p. An analogous procedure for the position x leads to the multiplicative operator

$$\tilde{x} = x \,. \tag{E.9}$$

Both the operators $\tilde{p}$ and $\tilde{x}$ act upon the classical Wigner function χ defined in (x, η)-space. Calculating the commutator

$$[\tilde{x}, \tilde{p}]\chi = (\tilde{x}\tilde{p} - \tilde{p}\tilde{x})\chi \,, \tag{E.10}$$

we find

$$[\tilde{x}, \tilde{p}]\chi = 0 \,, \tag{E.11}$$

which in turn leads to the relation between the fluctuations

$$\sigma_x \sigma_p \geq 0 \,. \tag{E.12}$$

Supposing that $\chi(x_1, x_2, t)$ in (E.7) is factorized as

$$\chi(x_1, x_2, t) = \phi^\dagger(x_2, t)\phi(x_1, t) \tag{E.13}$$

and expanding $\phi^\dagger(x + \eta\ell/2)$ and $\phi(x - \eta\ell/2)$, we can associate the operator

$$\hat{p} = -\mathrm{i}\ell\frac{\partial}{\partial x} \tag{E.14}$$

with p and the operator

$$\hat{x} = x \tag{E.15}$$

with x. These operators act upon the function ϕ. Hence

$$[\hat{x}, \hat{p}] = \mathrm{i}\ell \tag{E.16}$$

and

$$\sigma_x\sigma_p \geq \frac{\ell}{2} \, . \tag{E.17}$$

Taking

$$\ell \to \hbar \, , \tag{E.18}$$

we obtain the operator structure of quantum mechanics [209]. Therefore the transition from classical mechanics to quantum mechanics is a mathematically well-defined procedure, since both forms of mechanics are set into a common algebraic framework (q-numbers). Hence, before quantizing, the ordering rule is predetermined to be the symmetrical or Weyl rule, thus avoiding the well-known shortcomings inherent in the transition: (c-number) $\to$ (q-number).

References

1. M. Heidegger: *Die Frage nach dem Ding*, 3rd edn. (Max Niemeyer, Tübingen, 1987)
2. A. Einstein: *Out of my Later Years* (Wings Books, New York, 1956)
3. V.I. Arnold: *Ordinary Differential Equations* (MIT Press, Cambridge, Massachusetts, 1973)
4. M. Born: Vorhersagbarkeit in der klassischen Mechanik, Z. Phys. **153**, 372–388 (1958)
5. C. Lanczos: *The Variational Principles of Mechanics* (University of Toronto Press, Toronto, 1949)
6. J.C. Maxwell: *Matter and Motion* (Doubleday, New York, 1952)
7. W. Yourgrau, S. Mandelstam: *Variational Principles in Dynamics and Quantum Theory* (Dover, New York, 1968)
8. W. Heisenberg: Über quantentheoretische Umdeutung kinematischer und mechanistischer Beziehungen, Z. Phys. **33**, 879–893 (1925)
9. E. Schrödinger: Quantisierung als Eigenwertproblem, Ann. Phys. (Leipzig) **79**, 361–376 (1926)
10. P.A.M. Dirac: The fundamental equations of quantum mechanics, Proc. R. Soc. A **109**, 642–653 (1926)
11. R.P. Feynman: Space-time approach to non-relativistic quantum mechanics, Rev. Mod. Phys. **20**, 367–387 (1948)
12. W. Heisenberg: Über den anschaulichen Inhalt der quantentheoretischen Kinematik und Mechanik, Z. Phys. **43**, 172–198 (1927)
13. A.A. Grib, W.A. Rodrigues Jr.: *Nonlocality in Quantum Physics* (Kluwer, New York, 1999)
14. W. Heisenberg: *Der Teil und das Ganze* (Piper, München, 1969)
15. D. Bohm, B.J. Hiley: *The Undivided Universe: An Ontological Interpretation of Quantum Mechanics* (Routledge, London, 1993)
16. I. Prigogine: *Introduction to Thermodynamics of Irreversible Processes*, 3rd edn. (Interscience, New York, 1967)
17. G. Nicolis, I. Prigogine: *Self-Organization in Nonequilibrium Systems: From Dissipative Structures to Order Through Fluctuations* (Wiley, New York, 1977)
18. I. Prigogine: *From Being to Becoming* (Freeman, San Francisco, 1980)
19. I. Prigogine, I. Stengers: *Order out of Chaos* (Heinemann, London, 1984)

20. I. Prigogine: *La Fin des Certitudes: Temps, Chaos et les Lois de la Nature* (Édition Odile Jacob, Paris, 1996)
21. H. Haken: Cooperative phenomena in systems far from thermal equilibrium and in nonphysical systems, Rev. Mod. Phys. **47**, 67–121 (1975)
22. H. Haken: *Synergetics*, 3rd edn. (Springer, Berlin, 1983)
23. H. Haken: *Advanced Synergetics*, 2nd edn. (Springer, Berlin, 1987)
24. B.B. Mandelbrot: *The Fractal Geometry of Nature* (Freeman, New York, 1983)
25. E.N. Lorenz: Deterministic nonperiodic flow, J. Atm. Sci. **20**, 130–141 (1963)
26. E.N. Lorenz: *The Essence of Chaos* (University of Washington Press, Seatle, 1993)
27. D. Ruelle: *Chance and Chaos* (Princeton University Press, Princeton, New Jersey, 1991)
28. D. Ruelle: *Turbulence, Strange Attractors and Chaos* (World Scientific, Singapore, 1995)
29. A.J. Lichtenberg, M.A. Lieberman: *Regular and Chaotic Dynamics*, 2nd edn. (Springer, New York, 1992)
30. L. Arnold: *Random Dynamical Systems* (Springer, New York, 1998)
31. J. Ingen-Housz: Remarks on the use of the microscope. In: *Vermischte Schriften physisch-medizinschen Inhalts* (C.F. Wappler, Vienna, 1785)
32. P.W. van der Pas: The discovery of Brownian motion, Scientiae Historiae **13**, 27–35 (1971)
33. R. Brown: A brief account of microscopical observations, made in the months of June, July, and August, 1827 on the particles contained in the pollen of plants; and on the general existence of active molecules in organic and inorganic bodies, Edin. New Phil. J. **5**, 358–371 (1828)
34. R.M. Mazo: *Brownian Motion: Fluctuations, Dynamics and Applications* (Oxford University Press, Oxford, 2002)
35. J. Perrin: Mouvement brownien et réalité moléculaire, Ann. de Chim. et de Phys. **18**, 5–114 (1909)
36. A. Einstein: Über die von der molekularkinetischen Theorie der Wärme geforderte Bewegung von in ruhenden Flüssigkeiten suspendierten Teilchen, Ann. Phys. (Leipzig) **17**, 549–560 (1905)
37. A. Einstein: Zur Theorie der Brownschen Bewegung, Ann. Phys. (Leipzig) **19**, 371–381 (1906)
38. P. Langevin: Sur la théorie du mouvement brownien, C. R. Acad. Sci. Paris **146**, 530–533 (1908)
39. J. von Plato: *Creating Modern Probability* (Cambridge University Press, Cambridge, 1994)
40. A.O. Bolivar: Quantization of non-Hamiltonian physical systems, Phys. Rev. A **58**, 4330–4335 (1998)

41. A.O. Bolivar: Quantization and classical limit of a linearly damped particle, a van der Pol system and a Duffing system, Random Oper. and Stoch. Equ. **9**, 275–286 (2001)
42. A.O. Bolivar: The Wigner representation of classical mechanics, quantization and classical limit, Physica A **301**, 219–240 (2001)
43. A.O. Bolivar: Quantization of the anomalous Brownian motion, Phys. Lett. A **307**, 229–232 (2003)
44. A.O. Bolivar: Dynamical quantization and classical limit, Can. J. Phys. **81**, 663–673 (2003)
45. A.O. Bolivar: Classical limit of fermions in phase space, J. Math. Phys. **42**, 4020–4030 (2001)
46. A.O. Bolivar: Classical limit of bosons in phase space, Physica A **315**, 601–615 (2002)
47. A.O. Bolivar: The Bohm quantum potential and the classical limit of quantum mechanics, Can. J. Phys. **81**, 971–976 (2003)
48. C.C. Gillispie: *Pierre-Simon Laplace* (Princeton University Press, Princeton, 1997)
49. A. Kolmogorov: *Grundbegriffe der Wahrscheinlichkeitsrechnung* (Springer, Berlin, 1933)
50. A. Papoulis: *Probability, Random Variables, and Stochastic Processes* (McGraw-Hill, New York, 1991)
51. E. Cassirer: *Determinism and Indeterminism in Modern Physics* (Yale University Press, New Haven, 1956)
52. H. Cramér: *Mathematical Methods of Statistics* (Princeton University Press, Princeton, 1954)
53. R.L. Stratonovich: *Topics in the Theory of Random Noise*, Vol. 1 (Gordon and Breach, New York, 1963)
54. A. Kolmogorov: *Über die analytischen Methoden in der Wahrscheinlichkeitsrechnung*, Math. Ann. **104**, 414–458 (1931)
55. N.G. van Kampen: *Stochastic Processes in Physics and Chemistry* (North-Holland, Amsterdam, 1981)
56. G. Adomian: *Stochastic Systems* (Academic Press, New York, 1983)
57. T.T. Soong: *Random Differential Equations in Science and Engineering* (Academic Press, New York, 1973)
58. H. Bunke: *Gewöhnliche Differentialgleichungen mit zufälligen Parametern* (Akademie-Verlag, Berlin, 1972)
59. R.F. Pawula: Approximation of the linear Boltzmann equation by the Fokker–Planck equation, Phys. Rev. **162**, 186–188 (1967)
60. V. Volterra: *Theory of Functionals and of Integrals and Integrodifferential equations* (Dover, New York, 1959)
61. H. Cramér, M.R. Leadbetter: *Stationary and Related Stochastic Processes* (Wiley, New York, 1967)
62. M. Grigoriu: *Applied Non-Gaussian Processes* (Prentice Hall, Englewood Cliffs, NJ, 1995)

63. G. Samorodnitsky, M.S. Taqqu: *Stable Non-Gaussian Random Processes* (Chapman and Hall, New York, 1994)
64. D.T. Gillespie: *Markov Processes: An Introduction for Physical Scientists* (Academic Press, San Diego, 1992)
65. N.G. van Kampen: Remarks on non-Markov Processes, Braz. J. Phys. **28**, 90–96 (1998)
66. A.D. Fokker: Die mittlere Energie rotierender elektrischer Dipole im Strahlungsfeld, Ann. Phys. (Leipzig) **43**, 810–820 (1914)
67. M. Planck: Über einen Satz der statistischen Dynamik und seine Erweiterung in der Quantentheorie, Sitz. König. Preuss. Akad. Wiss. **1**, 324–341 (1917)
68. H. Risken: *The Fokker–Planck Equation: Methods of Solution and Applications*, 2nd edn. (Springer, Berlin, 1989)
69. C.W. Gardiner: *Handbook of Stochastic Methods: For Physics, Chemistry and the Natural Sciences*, 2nd edn. (Springer, Berlin, 1985)
70. J.G. Kirkwood: The statistical mechanical theory of transport processes, J. Chem. Phys. **14**, 180–201 (1946)
71. M.A. Olivares-Robles, L.S. García-Colín: Mesoscopic derivation of hyperbolic transport equations, Phys. Rev. E **50**, 2451–2457 (1994)
72. R. Fürth: Comments. In: A. Einstein, *Untersuchungen über die Theorie der Brownschen Bewegung*, Ostwalds Klassiker der Exakten Wissenschaften, Vol. 199, 3rd edn. (Deutsch, Frankfurt am Main, 1997)
73. L.F. Richardson: Atmospheric diffusion shown on a distance-neighbour graph, Proc. Roy. Soc. London A **110**, 709–737 (1926)
74. R. Zwanzig, M. Bixon: Hydrodynamic theory of the velocity correlation function, Phys. Rev. A **2**, 2005–2012 (1970)
75. T.S. Chow, J.J. Hermans: Effect of inertia on the Brownian motion of rigid particles in a viscous fluid, J. Chem. Phys. **56**, 3150–3154 (1972)
76. E.H. Hauge, A. Martin-Löf: Fluctuating hydrodynamics and Brownian motion, J. Stat. Phys. **7**, 259–281 (1973)
77. J.W. Dufty: Gaussian model for fluctuation of a Brownian particle, Phys. Fluids **17**, 328–333 (1974)
78. L. Reichl: Translational Brownian motion in a fluid with internal degrees of freedom, Phys. Rev. A **24**, 1609–1616 (1981)
79. G.L. de Haas-Lorentz: *Die Brownsche Bewegung und einige verwandte Erscheinungen* (Friedr. Vieweg und Sohn, Braunschweig, 1913)
80. M. Jo Nye: *Molecular Reality* (McDonald and American Elsevier, London New York, 1972)
81. J.C.M. Mombach, J.A. Glazier: Single cell motion in aggregates of embryonic cells, Phys. Rev. Lett. **76**, 3032–3035 (1996)
82. S.K. Banik, J.R. Chaudhuri, D.S. Ray: The generalized Kramers theory for nonequilibrium open one-dimensional systems, J. Chem. Phys. **112**, 8330–8337 (2000)

83. D. Ben-Avrahan, S. Havlin: *Diffusion and Reactions in Fractal and Disordered Systems* (Cambridge University Press, Cambridge, 2000)
84. S. Havlin, D. Ben-Avrahan: Diffusion in disordered media, Adv. Phys. **36**, 695–798 (1987)
85. J.P. Bouchad, A. Georges: Anomalous diffusion in disordered media: statistical mechanisms, models and physical applications, Phys. Rep. **195**, 127–293 (1990)
86. R. Metzler, J. Klafter: The random walk's guide to anomalous diffusion: A fractional dynamics approach, Phys. Rep. **339**, 1–77 (2000)
87. K. Kutner, A. Pekalski, and K. Sznajd-Weron (Eds.): *Anomalous Diffusion: From Basics to Applications Diffusion* (Springer, Berlin, 1999)
88. M.F. Schlesinger, G.M. Zaslavsky, and U. Frisch (Eds.): *Lèvy Flights and Related Topics in Physics* (Springer, Berlin, 1995)
89. A. Upadhyaya, J-P. Rieu, J.A. Glazier, Y. Sawada: Anomalous diffusion and non-Gaussian velocity distribution of Hydra cells in cellular aggregates, Physica A **293**, 549–558 (2001)
90. K.G. Wang, C.W. Lung: Long-time correlation effects and fractal Brownian motion, Phys. Lett. A **151**, 119–121 (1990)
91. K.G. Wang: Long-time-correlation effects and biased anomalous diffusion, Phys. Rev. A **45**, 833–837 (1992)
92. M. von Smoluchowski: Über Brownsche Molekularbewegung unter Einwirkung äusserer Kräfte und deren Zusammenhang mit der verallgemeinerten Diffusionsgleichung, Ann. Phys. (Leipzig) **48**, 1103–1112 (1915)
93. H.A. Kramers: Brownian motion in a field of force and the diffusion model of chemical reactions, Physica **7**, 284–304 (1940)
94. H.C. Brinkman: Brownian motion in a field of force and the diffusion theory of chemical reactions, Physica **22**, 29–34 (1956)
95. G. Wilemski: On the derivation of Smoluchowski equations with corrections in the classical theory of Brownian motion, J. Stat. Phys. **14**, 153–169 (1976)
96. L. Bocquet: From a stochastic to a microscopic approach to Brownian motion, Acta Phys. Polonica B **14**, 1551–1564 (1998)
97. S. Chandrasekhar: Stochastic problems in physics and astronomy, Rev. Mod. Phys. **15**, 1–89 (1943)
98. L.S. Ornstein: On the Brownian motion, Proc. Roy. Acad. Amsterdam **21**, 96–108 (1919)
99. G.E. Uhlenbeck, L.S. Ornstein: On the theory of the Brownian motion, Phys. Rev. **36**, 823–841 (1930)
100. M.C. Wang, G.E. Uhlenbeck: On the theory of the Brownian motion II, Rev. Mod. Phys. **17**, 323–342 (1945)
101. O. Klein: Zur statistische Theorie der Suspensionen und Lösungen, Arkiv f. Math. Astr. Phys. **16**, 1 (1922)

102. S.A. Adelman: Fokker–Planck equations for simple non-Markovian systems, J. Chem. Phys. **64**, 124–130 (1976)
103. R.M. Mazo: Aspects of the theory of Brownian motion. In: *Stochastic Processes in Nonequilibrium Systems*, ed. by L. Garrido, P. Seglar, and P.J. Shepherd (Springer, Berlin, 1978) pp. 53–81
104. P. Hänggi, F. Mojtabai: Thermally activated escape rate in presence of long-time memory, Phys. Rev. A **26**, 1168–1170 (1982)
105. J.M. Porrà, K-G. Wang, and J. Masoliver: Generalized Langevin equations: Anomalous diffusion and probability distributions, Phys. Rev. E **53**, 5872–5881 (1996)
106. K-G. Wang, J. Masoliver: Linear oscillators driven by Gaussian colored noise: crossovers and probability distributions, Physica A **231**, 615–630 (1996)
107. K-G. Wang, M. Tokuyama: Nonequilibrium statistical description of anomalous diffusion, Physica A **265**, 341–351 (1996)
108. M. Tabor: *Chaos and Integrability in Nonlinear Dynamics* (Wiley, New York, 1989)
109. V.I. Arnold: *Mathematical Methods of Classical Mechanics*, 2nd edn. (Springer, New York, 1989)
110. O. Penrose: Foundations of statistical mechanics, Rep. Prog. Phys. **42**, 1937–2006 (1979)
111. H. Spohn: Kinetic equations from Hamiltonian dynamics: Markovian limits, Rev. Mod. Phys. **52**, 569–615 (1980)
112. J.L. Lebowitz, E. Rubin: Dynamical study of Brownian motion, Phys. Rev. **131**, 2381–2396 (1963)
113. M. Toda: On the theory of the Brownian motion, J. Phys. Soc. Japan **13**, 1266–1280 (1958)
114. V.B. Magalinskii: Dynamical model in the theory of the Brownian motion, J. Exptl. Theor. Phys. **36**, 1942–1944 (1959)
115. P. Mazur, E. Braun: On the statistical mechanical theory of Brownian motion, Physica **30**, 1973–1988 (1964)
116. G.W. Ford, M. Kac, P. Mazur: Statistical mechanics of assemblies of coupled oscillators, J. Math. Phys. **6**, 504–515 (1965)
117. P. Ullersma: An exactly solvable model for Brownian motion I. Derivation of the Langevin equations, Physica **32**, 27–55 (1966)
118. P. Ullersma: An exactly solvable model for Brownian motion II. Derivation of the Fokker–Planck equation and the master equation, Physica **32**, 56–73 (1966)
119. P. Mazur, I. Oppenheim: Molecular theory of Brownian motion, Physica **50**, 241–258 (1970)
120. J.M. Deutch, R. Silbey: Exact generalized Langevin equation for a particle in a harmonic lattice, Phys. Rev. A **3**, 2049–2052 (1971)
121. R. Zwanzig: Nonlinear generalized Langevin equations, J. Stat. Phys. **9**, 215–220 (1973)

122. W. Bez: Microscopic preparation and macroscopic motion of a Brownian particle, Z. Physik B **39**, 319–325 (1980)
123. K. Lindenberg, B.J. West: *The Nonequilibrium Statistical Mechanics of Open and Closed Systems* (VCH Publisher, New York, 1990)
124. N.G. van Kampen, I. Oppenheim: Brownian motion as a problem of eliminating fast variables, Physica A **138**, 231–248 (1986)
125. Y. Zhang: Path-integral formalism for classical Brownian motion in a general environment, Phys. Rev. E **47**, 3745–3748 (1993)
126. R. Luzzi, A.R. Vasconcellos, J.G. Ramos: *Statistical Foundations of Irreversible Thermodynamics* (B.G. Teubner, Stuttgart, 2000)
127. R. Luzzi, A.R. Vasconcellos, J.G. Ramos: *Predictive Statistical Mechanics* (Kluwer, Dordrecht, 2002)
128. S.R.A. Salinas: *Introduction to Statistical Physics* (Springer, New York, 2001)
129. J.M. Blatt: An alternative approach to the ergodic problem, Prog. Theor. Phys. **22**, 745–756 (1959)
130. A.O. Caldeira, A.J. Leggett: Path integral approach to quantum Brownian motion, Physica A **121**, 587–616 (1983)
131. A.O. Caldeira, A.J. Leggett: Quantum tunnelling in a dissipative system, Ann. Phys. **149**, 374–456 (1983)
132. A.J. Leggett: Quantum tunnelling in the presence of an arbitrary linear dissipation mechanism, Phys. Rev. B **30**, 1208–1218 (1984)
133. C. Presilla, R. Onofrio, M. Patriarca: Classical and quantum measurements of position, J. Phys. A **30**, 7385–7411 (1997)
134. R.H. Koch, D.J. van Harlingen, J. Clarke: Quantum-noise theory for the resistively shunted Josephson junction, Phys. Rev. Lett. **45**, 2132–2135 (1980)
135. R.H. Koch, D.J. van Harlingen, J. Clarke: Measurements of quantum noise in resistively shunted Josephson junctions, Phys. Rev. B **26**, 74–87 (1982)
136. U. Weiss: *Quantum Dissipative Systems*, 2nd edn. (World Scientific, Singapore, 1999)
137. M. Planck: Über das Gesetz der Energieverteilung im Normalspektrum, Ann. Phys. (Leipzig) **4**, 553–566 (1901)
138. N. Bohr: On the constitution of atoms and molecules, Phil. Mag. **26**, 1–25 (1913)
139. W. Wilson: The quantum-theory of radiation and line spectra, Phil. Mag. **29**, 795–802 (1915)
140. M. Planck: Die physikalische Struktur des Phasenraumes, Ann. Phys. (Leipzig) **50**, 385–418 (1916)
141. A. Sommerfeld: Zur Quantentheorie der Spektrallinien, Ann. Phys. (Leipzig) **51**, 1–94 (1916)
142. K. Schwarzschild: Zur Quantenhypothese, Sitz. Preuss. Akad. Wiss. Berlin 548–568 (1916)

143. P.S. Epstein: Zur Theorie des Starkeffektes, Ann. Phys. (Leipzig) **50**, 489–520 (1916)
144. M. Jammer: *Conceptual Developments of Quantum Mechanics* (McGraw-Hill, New York, 1966)
145. A. Einstein: Zum Quantensatz von Sommerfeld und Epstein, Verh. deutsch. phys. Gesell. **19**, 82–92 (1917)
146. A.M. Ozorio de Almeida: *Hamiltonian Systems: Chaos and Quantization* (Cambridge University Press, Cambridge, 1988)
147. M.C. Gutzwiller: *Chaos in Classical and Quantum Mechanics* (Springer, Berlin, 1990)
148. M.N. Wise, D.C. Brock: The culture of quantum chaos, Stud. Hist. Phil. Mod. Phys. **29**, 369–389 (1998)
149. W. Heisenberg: Über quantentheoretische Kinematik und Mechanik, Math. Ann. **95**, 683–705 (1926)
150. M. Born, P. Jordan: Zur Quantenmechanik, Z. Phys. **34**, 858–888 (1925)
151. R. Benguria, M. Kac: Quantum Langevin equation, Phys. Rev. Lett. **46**, 1–3 (1981)
152. G.W. Ford, M. Kac: On the quantum Langevin equation, J. Stat. Phys. **46**, 803–810 (1987)
153. G.W. Ford, J.T. Lewis, R.F. O'Connell: Quantum Langevin equation, Phys. Rev. A **37**, 4419–4428 (1988)
154. C.W. Gardiner, P. Zoller: *Quantum Noise* (Springer, Berlin, 2000)
155. P.W. Milonni, W.A. Smith: Radiation reaction and vacuum fluctuations in spontaneous emission, Phys. Rev. A **11**, 814–824 (1975)
156. J. Audretsch, R. Müller: Spontaneous excitation of an accelerated atom: The contributions of vacuum fluctuations and radiation reaction, Phys. Rev. A **50**, 1755–1763 (1994)
157. D. Giulini, E. Joos, C. Kiefer, J. Kupsch, I.-O. Stamatescu, and H.D. Zeh: *Decoherence and the Appearance of a Classical World in Quantum Theory* (Springer, Berlin, 1996)
158. M. Schönberg: A non-linear generalization of the Schrödinger and Dirac equations, Nuovo Cimento **11**, 674–682 (1953)
159. H.-J. Wagner: Schrödinger quantization and variational principles in dissipative quantum theory, Z. Phys. B **93**, 261–273 (1994)
160. E. Schrödinger: Quantisierung als Eigenwertproblem, Ann. Phys. (Leipzig) **79**, 489–527 (1926)
161. B.R. Frieden: *Physics from Fisher Information: A Unification* (Cambridge University Press, Cambridge, 1998)
162. L. Pauling, E.B. Wilson Jr.: *Introduction to Quantum Mechanics* (McGraw-Hill, New York, 1935)
163. P.A.M. Dirac: *The Principles of Quantum Mechanics* (Clarendon, London, 1930)
164. W. Heisenberg: *The Physical Principles of Quantum Theory* (Doubleday, New York, 1930)

165. P.A.M. Dirac: Generalized Hamiltonian dynamics, Can. J. Phys. **2**, 129–149 (1950)
166. P.A.M. Dirac: *Lectures on Quantum Theory* (Yeshiva University, New York, 1964)
167. B. Podolski: Quantum-mechanical correct form of Hamiltonian function for conservative systems, Phys. Rev. **32**, 812–816 (1928)
168. J.R. Shewell: On the formation of quantum-mechanical operators, Am. J. Phys. **27**, 16–21 (1959)
169. A. Ashtekar: On the relation between classical and quantum observables, Comm. Math. Phys. **71**, 59–64 (1980)
170. I.R. Senitzky: Dissipation in quantum mechanics. The harmonic oscillator, Phys. Rev. **119**, 670–679 (1960)
171. A.O. Caldeira: Interference and coherent tunnelling in dissipative quantum systems, Helv. Phys. Acta **61**, 611–621 (1988)
172. Th.M. Nieuwenhuizen, A.E. Allahverdyan: Statistical thermodynamics of quantum Brownian motion: Construction of perpetuum mobile of the second kind, Phys. Rev. E **66**, 036102 (2002)
173. R.P. Feynman, F.L. Vernon Jr.: The theory of a general quantum system interacting with a linear dissipative system, Ann. Phys. (N.Y.) **24**, 118–173 (1963)
174. R.M. Cavalcanti: Wave function of a Brownian particle, Phys. Rev. E **458**, 6807–6809 (1998)
175. L.H. Yu, C.-Pu Sun: Evolution of the wave function in a dissipative system, Phys. Rev. A **49**, 592–595 (1994)
176. C.-Pu Sun, L.H. Yu: Exact dynamics of a quantum dissipative system in a constant external field, Phys. Rev. A **51**, 1845–1853 (1995)
177. L.H. Yu: Quantum tunneling in a dissipative system, Phys. Rev. A **54**, 3779–3782 (1996)
178. C.P. Sun, Y.B. Gao, H.F. Dong, S.R. Zhao: Partial factorization of wave functions for a quantum dissipative system, Phys. Rev. E **57**, 3900–3904 (1998)
179. P.A.M. Dirac: The Lagrangian in quantum mechanics, Phys. Z. Sowietunion **3**, 64–72 (1933)
180. R.L. Monaco, R.E. Lagos, W.A. Rodrigues Jr.: A Riemann integral approach to Feynman's path integral, Found. Phys. Lett. **8**, 365–373 (1995)
181. R.P. Feynman, A.R. Hibbs: *Quantum Mechanics and Path Integrals* (McGraw-Hill, New York, 1965)
182. L.S. Schulman: *Techniques and Applications of Path Integrals* (Wiley, New York, 1981)
183. C. Grosche, F. Steiner: *Handbook of Feynman Path Integrals* (Springer, Berlin, 1998)

184. M. Weyrauch, A.W. Schreiber: Comment on "Unique translation between Hamiltonian operators and functional integrals," Phys. Rev. Lett. **88**, 078901 (2002)
185. T. Gollisch, C. Wetterich: Unique translation between Hamiltonian operators and functional integrals, Phys. Rev. Lett. **86**, 1–5 (2000)
186. T.G. Dewey: Numerical mathematics of Feynman path integrals and the operator ordering problem, Phys. Rev. A **42**, 32–37 (1990)
187. F.J. Testa: Quantum operator ordering and the Feynman formulation, J. Math. Phys. **12**, 1471–1474 (1971)
188. A.K. Kapoor: Quantization in nonlinear coordinates via Hamiltonian path integrals, Phys. Rev. D **29**, 2339–2343 (1984)
189. S.S. Schweber: On Feynman quantization, J. Math. Phys. **3**, 831–842 (1962)
190. M.K. Ali: Quantization of the double pendulum, Can. J. Phys. **74**, 255–262 (1996)
191. K.H. Yeon, D.F. Walls, C.I. Um, T.F. George, L.N. Pandey: Quantum correspondence for linear canonical transformations on general Hamiltonian systems, Phys. Rev. A **58**, 1765–1774 (1998)
192. L. Cohen: Hamiltonian operators via Feynman path integrals, J. Math. Phys. **11**, 3296–3297 (1970)
193. E.H. Kerner, W.G. Sutcliffe: Unique Hamiltonian operators via Feynman path integrals, J. Math. Phys. **11**, 3296–3297 (1970)
194. I.W. Mayes, J.S. Dowker: Hamiltonian orderings and functional integrals, J. Math. Phys. **14**, 434–439 (1973)
195. M.M. Mizrahi: The Weyl correspondence and path integrals, J. Math. Phys. **16**, 2201–2206 (1975)
196. J.S. Dowker: Path integrals and ordering rules, J. Math. Phys. **17**, 1873–1874 (1976)
197. H.O. Girotti, T.J.M. Simões: Uniqueness of the functional approach, Phys. Rev. D **22**, 1385–1389 (1980)
198. E. Nelson: Derivation of the Schrödinger equation from Newtonian mechanics, Phys. Rev. **150**, 1079–1085 (1980)
199. T.G. Wallstrom: Inequivalence between the Schrödinger equation and the Madelung hydrodynamic equations, Phys. Rev. A **49**, 1613–1617 (1994)
200. A.F. Kracklauer: Comment on derivation of the Schrödinger equation from Newtonian mechanics, Phys. Rev. D **10**, 1358–1360 (1974)
201. H. Grabert, P. Hänggi, P. Talkner: Is quantum mechanics equivalent to a classical stochastic process? Phys. Rev. A **19**, 2440–2445 (1979)
202. E. Nelson: *Quantum Fluctuations* (Princeton University Press, Princeton, New Jersey, 1985)
203. L.S.F. Olavo: Foundations of quantum mechanics: Connection with stochastic processes, Phys. Rev. A **61**, 052109 (2000)

204. L. de la Peña, A.M. Cetto: *The Quantum Dice: An Introduction to Stochastic Electrodynamics* (Kluwer, Dordrecht, 1996)
205. L.S.F. Olavo: Foundations of quantum mechanics (II): Equilibrium, Bohr–Sommerfeld rules and duality, Physica A **271**, 260–302 (1999)
206. L.S.F. Olavo, A.F. Bakuzis, R.Q. Amilcar: Generalized Schrödinger equation using Tsallis entropy, Physica A **271**, 303–323 (1999)
207. L.S.F. Olavo: Possible physical meaning of the Tsallis entropy parameter, Phys. Rev. E **64**, 036125 (2001)
208. C. Tsallis: Nonextensive statistics: Theoretical, experimental and computational evidences and connections, Braz. J. Phys. **29**, 1–35 (1999)
209. L.S.F. Olavo: Foundations of quantum mechanics: Non-relativistic theory, Physica A **262**, 197–214 (1999)
210. J-M. Lévy-Leblond, F. Balibar: *Quantics: Rudiments of Quantum Physics* (North-Holland, Amsterdam, 1990)
211. M. Patriarca: Statistical correlations in the oscillator model of quantum dissipative systems, Nuovo Cimento B **111**, 61–72 (1996)
212. C. Morais Smith, A.O. Caldeira: Application of the generalized Feynman–Vernon approach to a simple system: The damped harmonic oscillator, Phys. Rev. A **41**, 3103–3115 (1990)
213. P.M.V.B. Barone, A.O. Caldeira: Quantum mechanics of radiation damping, Phys. Rev. A **43**, 57–63 (1991)
214. P.M.V.B. Barone, C. Morais Smith, D.S. Galvão: Numerical study of transport in a dissipative medium, Phys. Rev. A **45**, 3592–3595 (1992)
215. A.C.R. Mendes, F.I. Takakura: Radiation damping in real time, Phys. Rev. E **64**, 056501 (2001)
216. M.R. da Costa, A.O. Caldeira, S.M. Dutra, H. Westfahl Jr.: Exact diagonalization of two quantum models for the damped harmonic oscillator, Phys. Rev. A **61**, 022107 (2000)
217. C. Morais Smith, A.O. Caldeira: Generalized Feynman–Vernon approach to dissipative quantum systems, Phys. Rev. A **36**, 3509–3511 (1987)
218. R. Bedau: Dissipation und Dekohärenz in gekoppelten Qubits. Diplomarbeit, Universität Stuttgart, Stuttgart (2003)
219. L.D. Romero, J.P. Paz: Decoherence and initial correlations in quantum Brownian motion, Phys. Rev. A **55**, 4070–4083 (1997)
220. R. Karrlein, H. Grabert: Exact time evolution and master equations for the damped harmonic oscillator, Phys. Rev. E **55**, 153–164 (1997)
221. A.O. Caldeira, A.H.C. Neto, T.O. Carvalho: Dissipative quantum systems modeled by a two-level-reservoir coupling, Phys. Rev. B **48**, 13974–13976 (1993)
222. P.C. Marques, A.H.C. Neto: Master equation for a particle coupled to a two-level-reservoir, Phys. Rev. B **52**, 10693–10696 (1995)
223. J.T. Stockburger, H. Grabert: Exact *c*-number representation of non-Markovian quantum dissipation, Phys. Rev. Lett. **88**, 170407 (2002)

224. S. Stenholm: Quantum theory of linear friction, Braz. J. Phys. **27**, 214–237 (1997)
225. V. Ambegaokar: Quantum Brownian motion and its classical limit, Ber. Bunsenges. Phys. Chem. **95**, 400–404 (1991)
226. S. Gao: Dissipative quantum dynamics with a Lindblad functional, Phys. Rev. Lett. **79**, 3101–3104 (1997)
227. K. Jakobs, I. Tittonen, H.M. Wiseman, S. Schiller: Quantum noise in the position measurement of a cavity mirror undergoing Brownian motion, Phys. Rev. A **60**, 538–548 (1999)
228. A.O. Caldeira, H.A. Cerdeira, R. Ramaswamy: Limits of weak damping of a quantum harmonic oscillator, Phys. Rev. A **40**, 3438–3440 (1989)
229. E. Madelung: Quantentheorie in hydrodynamischer Form, Z. Phys. **40**, 322–326 (1927)
230. P.R. Holland: *The Quantum Theory of Motion: An Account of the de Broglie–Bohm Causal Interpretation of Quantum Mechanics* (Cambridge University Press, Cambridge, 1993)
231. G. Lindblad: On the generators of quantum dynamical semigroups, Comm. Math. Phys. **48**, 119–130 (1976)
232. J.G. Peixoto de Faria, M.C. Nemes: Phenomenological criteria for the validity of quantum Markovian equations, J. Phys. A **31**, 7095–7103 (1998)
233. H.M. Wiseman, W.J. Munro: Comment on "Dissipative quantum dynamics with a Lindblad functional," Phys. Rev. Lett. **80**, 5702 (1998)
234. S. Gao: Phys. Rev. Lett. **80**, 5703 (1998)
235. G.W. Ford, R.F. O'Connell: Comment on "Dissipative quantum dynamics with a Lindblad functional," Phys. Rev. Lett. **82**, 3376 (1999)
236. S. Gao: Phys. Rev. Lett. **82**, 3377 (1998)
237. B. Vacchini: Completely positive quantum dissipation, Phys. Rev. Lett. **84**, 1374–1377 (2000)
238. R.F. O'Connell: Comment on "Completely positive quantum dissipation," Phys. Rev. Lett. **87**, 028901 (2001)
239. B. Vacchini: Phys. Rev. Lett. **87**, 028902 (2000)
240. F.J. Kennedy Jr., E.H. Kerner: Note on the inequivalence of classical and quantum Hamiltonians, Am. J. Phys. **33**, 463–466 (1965)
241. D. Park: *Classical Dynamics and its Quantum Analogues*, 2nd edn. (Springer, Berlin, Heidelberg, 1990)
242. G.G. Cabrera, M. Kiwi: Large quantum-number states and the correspondence principle, Phys. Rev. A **36**, 2995–2998 (1987)
243. B. Gao: Breakdown of Bohr's correspondence principle, Phys. Rev. Lett. **83**, 4225–4228 (1999)
244. C. Eltschka, H. Friedrich, M.J. Moritz: Comment on "Breakdown of Bohr's correspondence principle," Phys. Rev. Lett. **86**, 2693 (2001)
245. C. Boisseau, E. Audouard, J. Vigué: Comment on "Breakdown of Bohr's correspondence principle," Phys. Rev. Lett. **86**, 2694 (2001)

246. P. Ehrenfest: Bemerkung über die angenärte Gültigkeit der klassischen Mechanik innerhalb der Quantenmechanik, Z. Phys. **45**, 455–457 (1927)
247. L.E. Ballentine, Y. Yang, J.P. Zibin: Inadequacy of Ehrenfest's theorem to characterize the classical regime, Phys. Rev. A **50**, 2854–2859 (1994)
248. G. Wenzel: Eine Verallgemeinerung der Quantenbedingungen für die Zwecke der Wellenmechanik, Z. Phys. **38**, 518–529 (1926)
249. H. Kramers: Wellenmechanik und halbzahlige Quantisierung, Z. Phys. **39**, 828–840 (1926)
250. L. Brilloin: Remarques sur la mécanique ondulatoire, J. Phys. et Rad. **6**, 333–363 (1926)
251. Ph. Blanchard, D. Giulini, E. Joos, C. Kiefer, I.-O.Stamatescu (Eds.): *Decoherence: Theoretical, Experimental, and Conceptual Problems* (Springer, Berlin Heidelberg, 2000)
252. W. Zurek: Decoherence, einselection, and the quantum origins of the classical, Rev. Mod. Phys. **75**, 715–775 (2003)
253. A.O. Caldeira, A.J. Leggett: Influence of damping on quantum interference: An exactly soluble model, Phys. Rev. A **31**, 1059–1066 (1985)
254. R. Omnes: Consistent interpretations of quantum mechanics, Rev. Mod. Phys. **64**, 339–382 (1992)
255. A.J. Leggett: Testing the limits of quantum mechanics: Motivation, state of play, prospects, J. Phys.: Condens. Matter **14**, R415–R451 (2002)
256. P. Busch, P.J. Lahti, P. Mittelstaedt: *The Quantum Theory of Measurement* (Springer, Berlin Heidelberg, 1991)
257. D. Bohm, B.J. Hiley: Unbroken quantum realism, from microscopic to macroscopic levels, Phys. Rev. Lett. **55**, 2511–2514 (1985)
258. A.J. Makowski: Exact classical limit of quantum mechanics: Central potentials and specific states, Phys. Rev. A **65**, 032103 (2002)
259. A.J. Makowski, K.J. Górska: Bohr's correspondence principle: The cases for which it is exact, Phys. Rev. A **66**, 062103 (2002)
260. P.R. Holland: Is quantum mechanics universal? In: *Bohmian Mechanics and Quantum Theory: An Appraisal*, ed. by J.T. Cushing, A. Fine, S. Goldstein (Kluwer, Dordrecht, 1996) pp. 99–110
261. J.T. Cushing: Bohmian mechanics and chaos. In: *From Physics to Philosophy*, ed. by J. Butterfield, C. Pagonis (Cambridge University Press, Cambridge, 1999) pp. 90–107
262. E.P. Wigner: On the quantum correction for thermodynamic equilibrium, Phys. Rev. **40**, 749–759 (1932)
263. E.J. Heller: Wigner phase space method: Analysis for semiclassical applications, J. Chem. Phys. **65**, 1289–1298 (1976)
264. J.G. Kirkwood: Quantum statistics of almost classical assemblies, Phys. Rev. **44**, 31–37 (1933)
265. R.F. O'Connel, E.P. Wigner: Some properties of a non-negative quantum-mechanical distribution function, Phys. Lett. A **85**, 121–126 (1981)

266. A. Einstein: Elementare Überlegungen zur Interpretation der Grundlagen der Quanten-Mechanik. In: *Scientific Papers Presented to Max Born on his Retirement from the Tait Chair of Natural Philosophy in the University of Edinburgh* (Oliver and Boyd, Edinburgh, 1953) pp. 33–40
267. A. Einstein: In: *The Born–Einstein Letters* (Macmillan, London, 1971)
268. W. Pauli: In: *The Born–Einstein Letters* (Macmillan, London, 1971)
269. M. Born: Continuity, determinism and reality, Det Kong. Dansk. Vidensk. Sels. Mate-fys. Medd. **30**, 1–26 (1955)
270. L.E. Ballentine: *Quantum Mechanics* (World Scientific, New York, 1998)
271. H.J. Carmichael: *Statistical methods in quantum optics I. Master equations and Fokker–Planck equations* (Springer, Berlin Heidelberg, 1999)
272. G. Mahler, V.A. Weberruss: *Quantum Networks: Dynamics of Open Nanostructures*, 2nd edn. (Springer, Berlin Heidelberg, 1998)

Index